First Steps in Real Analysis

First Steps in Real Analysis

Jonathan Britt
*Former Head of School of Computer Science and Mathematics,
University of Portsmouth*

Copyright ©Jonathan Britt 2014

First published in 2014 by Camera Caliraya

Camera Caliraya
Caliraya Manor
West Talaongan
Cavinti
Laguna
Philippines

All rights reserved. No part of this publication may be reproduced, stored in a retrieval system, or transmitted, in any form or by any means, electronic, mechanical, photocopying or otherwise, without the prior permission of the publisher, Camera Caliraya.

The Author has asserted his right under the Copyright Designs and Patents Act, 1988, to be identified as the Author of this Work.

Printed by CreateSpace, An Amazon.com Company

Preface

Some books just take a long time to write!!

This book stems from 1966/67 and it only 42 years later that it starts to get written. In fact the origins of this book go back to the first and second terms of the academic year 1966-67 at the University of Warwick. As a first year mathematics undergraduate there, we were given a two term course called "Differential Equations and Analysis" by Dr. (later Prof.) Rolph Schwarzenberger, an extremely gifted communicator.

The course was designed to be an eye-opener for us. The first part, covering Differential Equations, eschewed the "categorise and solve" approach that we had met in our previous studies and instead looked at how to use a differential equation to define a vector field and hence produce a solution curve by semi-graphical means.

The second part, covering Analysis, was just as revolutionary. Telling us that the "traditional' approach to teaching Analysis was first limits, next continuity then differentiability and integrabilty, (all defined in terms of limits), we were to be exposed to a different order, continuity first, then differentiability and limits to come only later. It was hoped that some of us would go on as mathematics teachers to use this new approach ourselves. (The reasons given for adopting this approach I will explain in the Introduction below.)

There were of course, no textbooks available which covered the material in the way we were dealing with it. In fact this was true for many, if not all of our courses. After a year Chapman and Hall began publishing a series of textbooks many of which were written by the Warwick mathematics lecturers and covering the contents of their courses. Rolph Schwarzenberger wrote one such book, published in the Chapman and Hall series, which covered the Differential Equations part of the course. [Sch69]

The Analysis part was covered by another book in the Chapman and Hall series, A Preliminary Course in Analysis, written by two lecturers from the University of Hull R.M. F. Moss and G.T. Roberts. [MR68] Unfortunately this book suffered from the disadvantage of having a lot of notation, much of which was non-standard. This made it difficult to read or use for first year undergraduates. Meanwhile the course at Warwick was taken over and taught by another equally gifted teacher, David Fowler.

By 1973 I was teaching first year Analysis myself to part-time degree students at Southampton College of Technology (now Southampton Solent University). I was using the same approach we had met at Warwick. The lack of a textbook was a serious drawback for the kind of students we had and, as explained above, Moss and Roberts, which was the only one available using this approach was too difficult. Then David Fowler wrote and published an amazing book entitled "Introducing Real Analysis" in the Transworld Student Library series. [Fow73] Unlike all other Analysis textbooks, this one was written in English. David Fowler abandoned the traditional Definition, Lemma, Theorem approach and wrote instead in connected narrative prose. The result was an easy to read and easy to understand book. Unfortunately it only covered Continuity and Differentiability, though it did contain a chapter indicating how the book could be continued. Because it only covered what one would consider as the first half of an Analysis course, I decided to try and write a sequel myself. At the same time I wanted to defer the idea of a limit as much as possible and so tried to utilise an approach to integration that would not use limits. Hence I came upon the idea of defining the integral over an interval in terms of the average of a function's value on that interval. (Again more about this in the Introduction below).

The sequel I wrote, entitled "Analysis- The Second Step' was written in a style and manner that was a conscious imitation of David Fowler's. It covered Integration, Sequences and Series, Power Series and the Elementary Functions. The book was only published within the institution for my own students and as my academic career changed from Mathematics to Computing, it lay forgotten until my retirement. It was then that I came across my own copy of my sequel and thought about revising it. I then realised that David Fowler's book was long out of print, so a sequel to that work made little sense. So the decision was made to try and write a text book to cover all of the contents of a first year Analysis course, following that Warwick approach.

I decided not to continue the "narrative prose" style of writing, both because I was clearly not as good at it as David Fowler had been and also because I thought that first year mathematics undergraduates needed to get to grips with the formal style of writing mathematics as soon as they could. Hence this book follows a conventional "Definition, Lemma, Theorem" style.

My main problem was to compose material in parallel to David Fowler's work, that is to cover Continuity and Differentiability, without slavishly following him. In fact this was easier than I had originally thought, partly because I came across a completely new (to me) definition of continuity for mapping from one topological space to another, in a book by K.G. Binmore, [Bin81] which I was able to use and convert into a real analysis setting and partly too because I had already written, some years ago a programmed text on Differentiability, which I was able to re-use.

Of course there will remain some similarities, most based upon the original lecture notes of Rolph Schwarzenberger, but I have given all the material my own flavouring!

Jonathan Britt
Cavinti
Laguna
Philippines
January 2010

Contents

Preface	iii
Introduction	1

1 Sets, Numbers and Functions — 5
- 1.1 Objectives ... 5
- 1.2 New terms and symbols introduced ... 6
- 1.3 Sets and Relations ... 7
- 1.4 Numbers: Whole, Rational and Real ... 13
- 1.5 Some Properties of the Reals ... 17
- 1.6 Definition and Simple Properties of Functions ... 24
- 1.7 Further Properties of Functions ... 29
- 1.8 Exercises ... 33

2 Continuity — 39
- 2.1 Objectives ... 39
- 2.2 New terms and symbols introduced ... 39
- 2.3 What does Continuous mean? ... 39
- 2.4 Properties of Continuous Functions ... 43
- 2.5 Completeness Properties of Continuous Functions ... 53
- 2.6 Exercises ... 60

3 Differentiation - Basic Definitions — 63
- 3.1 Objectives ... 63
- 3.2 New terms and symbols introduced ... 63
- 3.3 What is differentiation? ... 63
- 3.4 Exercises ... 73

4 Elementary properties of differentiable functions:1 — 75
- 4.1 Objectives ... 75
- 4.2 The sum and product of two functions ... 75
- 4.3 The composition of two functions ... 77
- 4.4 Exercises ... 79

5 Elementary properties of differentiable functions:2 — 81
- 5.1 Objectives ... 81
- 5.2 The reciprocal of a function and quotient of two functions ... 81
- 5.3 The inverse of a function ... 84
- 5.4 Summary of results from Chapters 4 and 5 ... 86

5.5	Exercises	86

6 Significance of the Derivative 87
6.1 Objectives 87
6.2 New Terms introduced 87
6.3 Critical Points and Critical Values 87
6.4 The Mean Value Theorem 92
6.5 Exercises 94

7 Averages 97
7.1 Objectives 97
7.2 New Terms introduced 97
7.3 Introduction 97
7.4 Definition of an Average 98
7.5 Uniqueness of an Average 102
7.6 What functions are Averageable? 105
7.7 Exercises 111

8 Step Functions 113
8.1 Objectives 113
8.2 New Terms introduced 113
8.3 Introduction 113
8.4 Construction of the Lower Step Function 115
8.5 Definition of the Average of a Continuous Function over an interval 116
8.6 Proof that this is an Average 117
8.7 Exercises 120

9 The Integral of a Function 121
9.1 Objectives 121
9.2 New Terms introduced 121
9.3 Introduction 121
9.4 Elementary properties of the definite integral 122
9.5 The relationship between differentiation and integration 123
9.6 Further properties of the integral 124
9.7 Exercises 126

10 Sequences and Series 131
10.1 Objectives 131
10.2 New Terms introduced 131
10.3 Introduction 131
10.4 Sequences and Limits 132
10.5 Series 133

10.6	Convergence of Series	136
10.7	Exercises	141

11 Power Series   145
11.1	Objectives	145
11.2	New Terms introduced	145
11.3	Introduction and Definition	145
11.4	Convergence	146
11.5	Continuity of Power Series	149
11.6	Differentiability of Power Series	151
11.7	Integration of Power Series	155
11.8	Exercises	156

12 The Elementary Functions   159
12.1	Objectives	159
12.2	New Terms introduced	159
12.3	Definitions	159
12.4	Some Properties of the Exponential Function	161
12.5	Some Properties of the Trigonometric Functions	162
12.6	Pi, π, 3.14159...	163
12.7	Exercises	168

Index   173

CONTENTS

Introduction

This book includes all the material that you would expect to find in a first year undergraduate course in Mathematical Analysis. That is, it covers, numbers, particularly the real numbers, functions (real valued of one variable), continuity, differentiability and integrability of such functions, sequences, series and power series.

However it approaches these topics using an approach that differs considerably from the "traditional" one, by which we mean one where the emphasis is placed upon an early definition of a limit which is then used in the subsequent definitions of continuity and differentiability.

In this book we take a more geometric road to defining those two concepts and leave the idea of a limit for much later. The reasons for doing this are two-fold. Firstly, the definition of the limit is conceptually difficult for a student who is only just beginning their study of rigorous pure mathematics. It seems to bear little or no relation to any intuitive ideas they may have held previously. In contradistinction to that, the concept of continuity is actually a very easy and intuitive one to understand. Even though some work needs to be done to move from that intuitive idea to a rigorous definition, nevertheless that move is much easier to explain and for the student, easier to understand than the traditional imposition of the concept of limit.

Secondly, the definition of the derivative, using the limit concept, only makes sense for functions from the reals to themselves; as soon as a student meets functions from $\mathbb{R}^2$ to itself, for example, a whole new definition has to be written. In order to show that such a definition is essentially the same as that used for real valued functions of one variable, it makes sense to use a definition that can be easily generalized in an obvious manner.

So the order of introducing those key concepts in this book is " Continuity - Differentiability - Limits". Furthermore, when it comes to integration of functions from the reals to themselves, a route is taken which is slightly different again from the more traditional one, and also avoids the direct use of the limit concept. That concept is replaced by the ideas of a greatest lower bound and least upper bound, both of which are introduced earlier in relation to talking about the

1

real numbers. In this way, the actual definition of a limit is deferred until very late in the text, where it is introduced and dealt with in the context of sequences of numbers.

The book is intended to be read in the strict order in which it is written, each chapter depends upon and builds on the preceding ones. The first chapter is intended to be a revision chapter, recapping the important results from set theory which are going to be needed and then going on to discuss the ideas of numbers, natural, integer, rational and real, before introducing the idea of a function and its elementary properties. The chapter is therefore written in a fairly terse manner, since it is assumed that the student should have met most if not all of those ideas elsewhere. Although formal definitions of the natural numbers, integers and rationals are given, no definition is provided for the reals. (The concept of Cauchy sequences of rationals is not used anywhere in the book). Instead the reals are characterized by their properties, in particular the existence of least upper bounds and greatest lower bounds. The material on functions is taken at a slower pace, as it is not assumed that the student has met these ideas before.

Chapter Two is devoted to Continuity of Functions. Some time is taken to define the concept and to explain how we arrive at that definition. Then all of the standard results, continuity of the sum of two continuous functions, and the product of two continuous functions, etc. etc. are proved. Finally the Intermediate Value Theorem is proved.

The next four chapters are devoted to the concept of differentiation. In the first of these chapters, time is taken to explain the concept before the actual definition is made. The next two chapters look at the properties of and rules for differentiation, while the last chapter in this set examines critical points and values, ending with the Mean Value Theorem.

The following three chapters are concerned with the integral of a function. The approach here is unusual. Firstly the concept of the average of a function over an interval is introduced and we show how it can be calculated, (provided the function concerned is the derivative of some other function), by using the Mean Value Theorem. The concept is then extended to other functions, those which are not the derivative of a function but are still piecewise continuous, by constructing lower step functions for those functions. Along the way the linearity properties of the average are established.

Finally the average is converted into a definite integral by multiplying it by the length of the interval. All the standard properties of the integral, linearity, the first and second Fundamental Theorems of Calculus and integration by substitution and by parts are derived.

In Chapter 10 attention is turned to sequencies and series of real numbers. At last the limit of a sequence is formally defined and the definition of a convergent series is also met. The treatment is brief and there is much else that could have been included here. The criterion used for inclusion of material here was simply that which would be useful in the final two chapters.

The next chapter, Chapter 11, deals with power series. After defining what a power series is, the conditions for convergence are investigated. We then show how a power series defines a continuous function wherever it converges. It is then shown how to differentiate and also integrate a power series term by term on that convergence set.

The very last chapter defines the exponential, logarithmic and trigonometric functions. This definition is made by using convergent power series. Following the definitions, the basic and familiar properties of these functions are proved, and the book concludes with showing a way (not the best way though) of calculating π.

The structure of each chapter is the same. The chapter begins with a brief statement of its objectives, followed by a glossary of new terms and symbols that are going to be introduced. Within the main body of each chapter all the major results are proved. However where there is a degree of repetition in a proof, or one proof is essentially the same as another, part or all of that proof may be left to the reader as an exercise. After each chapter is a section of exercises, which includes all of those "left to the reader" exercises from the chapter as well as some new exercises. Some of these are intended to consolidate the ideas within the chapter and occasionally some of them provide a way of seeing further, of building upon the given material to show other directions within the subject which were not included in the main text.

Of course the exercises are essential! Doing them will help the student gain a familiarity with the style of argument used in Analysis as well as deepening their understanding of the concepts and techniques discussed.

Solutions to the exercises are not provided, and I recognise that this is a controversial decision. My rationale is that if solutions were given, students may be tempted to look at such solutions too quickly and not try to reason out the answers for themselves. However I appreciate the contrary view that without such solutions, it can be hard for students to know if they are on then right track and if they are understanding what is being taught. At some later stage then solutions will be made available, probably on the web and possibly password protected and restricted to instructors rather than students! The instructors can then make the decision as to what would be best for their students.

Chapter 1

Sets, Numbers and Functions

1.1. Objectives

This chapter serves a twofold purpose. Firstly it is a revision chapter. It revises those ideas from Set Theory that we are going to need as well as some ideas about numbers and number systems which you may well have met before. It also revises the concepts of a function and its graph.

Secondly it formalises some of those elementary ideas and also develops them a little further.

So by the end of the chapter you will know what is meant by all of the terms listed in Section 1.2 and be able to use those terms in exercises and proofs.

Now there is a very long list of terms in that section, don't be put off by that, the ideas are mostly ones you will have met before and are not complicated. Just to give you a little guidance, the most important new terms that you should concentrate on include, *supremum, infimum, open and closed intervals, neighbourhood of a point, domain, range and inverse of a function.*

1.2. New terms and symbols introduced

set	A
members of a set	$\{a, b, c, \ldots\}$
subset of a set	
union of two sets	$A \cup B$
intersection of two sets	$A \cap B$
set difference of two sets	$A \setminus B$
empty set	$\emptyset$
cartesian product of two sets	$A \times B$
ordered pairs of elements of sets	(a, b)
relation on a set	ρ
transitive relation	
Natural Numbers	$\mathbb{N}$
Integers	$\mathbb{Z}$
Rational Numbers	$\mathbb{Q}$
Real Numbers	$\mathbb{R}$
Complex Numbers	$\mathbb{C}$
field	$\mathbb{F}$
ordered set	
ordered field	
supremum or least upper bound	sup
infimum or greatest lower bound	inf
complete ordered field	
interval	I
bounded interval	
unbounded interval	
open interval	(a, b)
closed interval	$[a, b]$
neighbourhood of a point x	N_x
boundary points of an interval	
modulus or absolute value of a real number x	$|x|$
triangle inequality	

contiguous sets
function f
graph of a function
domain of a function
image of a set A under a function $\qquad$ $f(A)$
range of a function
inverse of a function f $\qquad$ f^{-1}
inverse image of a set A under a function $\qquad$ $f^{-1}(A)$
polynomial function
rational function
one to one function
onto function

1.3. Sets and Relations

These days most people will know what a set is and the various things that can be done to and with sets, namely, constructing subsets of a set, forming the union, intersection, and set difference of two sets. So I am going to define all these things fairly quickly assuming that I am just refreshing your memory! If that is not the case, then please stop here and read a book that is explicitly about set algebra before going any further!

Definition 1.3.1. *A* **set** *is a collection of distinct objects, which are called the* **elements** *or* **members** *of the set.*

Although nearly all of the sets we are going to meet have numbers as their elements, you should be aware that any object at all can be a member of a set.
I will refer to a set by using a capital letter like, A, and if I need to write the members of a set explicitly I shall do so by listing them within curly brackets, so, for example, the set of vowels in English is $\{a, e, i, o, u\}$.

Definition 1.3.2. *If A is a set then by a **subset** of A we mean any set B constructed so that all the members of B are members of A. We say that B is contained in A.*

If A and B are two sets then we define $A \cup B$, the **union** of A and B by:
$A \cup B =$ all the elements of A and all the elements of B, (but if any element is in both A and B, it only gets included once!)

If A and B are two sets then we define $A \cap B$, the **intersection** of A and B by:
$A \cap B =$ all the elements of A that are also elements of B

If A and B are two sets then we define $A \setminus B$, the **set difference** of A and B by:
$A \setminus B =$ all the elements of A with any element that is in both A and B removed

Remember that a set A is always a subset of itself and also that there is a special set called the empty set, a set with no members at all (normally written as $\emptyset$), which is a subset of *every* set.
I hope that is all clear. Just to make sure, here are some examples.

Examples 1.3.1
1. Let $A = \{a, e, i, o, u\}$, the set of English vowels. Then A has the following subsets:

The empty set, $\emptyset$, with no members
The five sets $\{a\}, \{e\}, \{i\}, \{o\}, \{u\}$, all with one member
The ten sets $\{a, e\}, \{a, i\}, \{a, o\}, \{a, u\}, \{e, i\}, \{e, o\}, \{e, u\}, \{i, o\}, \{i, u\}, \{o, u\}$, all with two members
The ten sets $\{a, e, i\}, \{a, e, o\}, \{a, e, u\}, \{a, i, o\}, \{a, i, u\}, \{a, o, u\}, \{e, i, o\}, \{e, i, u\}, \{e, o, u\}, \{i, o, u\}$, all with three members
The five sets $\{a, e, i, o\}, \{a, e, i, u\}, \{a, e, o, u\}, \{a, i, o, u\}, \{e, i, o, u\}$, all with four members
And lastly A itself, namely $\{a, e, i, o, u\}$, with five members.
So A has 32 subsets in all!

2. Let $A = \{1, 2, 3, 4, 5, 6\}$ and $B = \{2, 6, 8, 10\}$ then:

$$A \cup B = \{1, 2, 3, 4, 5, 6, 8, 10\}$$

and
$$A \cap B = \{2, 6\}$$

and
$$A \setminus B = \{1, 3, 4, 5\}$$

3. Let $A = \{red, orange, yellow, green, blue, indigo, violet\}$ and $B = \{cyan, magenta, yellow, green\}$, then:

$$A \cup B = \{red, orange, yellow, green, blue, indigo, violet, cyan, magenta\}$$

and
$$A \cap B = \{yellow, green\}$$

and
$$A \setminus B = \{red, orange, blue, indigo, violet\}$$

Hopefully all those definitions are clear so we can go on to the next few definitions that you may not have met before.

Ordered pairs and the cartesian product of two sets

This is a very important idea so we will take it a little more slowly! Suppose we have two sets, A a set of children and B a set of presents, then we could pair each child with any of the presents. Such a pairing is known as an ordered pair.

So, for example, if
$A = \{Tony, Imelda, Sacha, Louisa\}$ and $B = \{balloon, chocolates, book\}$
we can form the pairs:
(Tony, balloon), (Tony, chocolates), (Tony, book), (Imelda, balloon), (Imelda, chocolates), (Imelda, book), (Sasha, balloon), (Sasha, chocolates), (Sasha, book), (Louisa, balloon), (Louisa, chocolates), (Louisa, book). Notice we have 12 ($= 4 \times 3$) pairings.

Now let's formalise this:

Definition 1.3.3. *If A and B are two sets then we define $A \times B$, the **cartesian product** of A and B by:*
$A \times B =$ *all the ordered pairs of elements with the first element in the pair taken from A and the second element in the pair taken from B.*

Examples 1.3.2
1. Let $A = \{1, 2, 3, 4, 5\}$ and $B = \{a, b, c, d\}$, then:

$$\begin{aligned} A \times B = \{&(1,a),(1,b),(1,c),(1,d),(2,a),(2,b),(2,c),(2,d),\\ &(3,a),(3,b),(3,c),(3,d),(4,a),(4,b),(4,c),(4,d),\\ &(5,a),(5,b),(5,c),(5,d)\} \end{aligned}$$

Note that $A \times B$ has $5 \times 4 = 20$ members.

2. Let $\mathbb{Z} = \{\ldots, -4, -3, -2, -1, 0, 1, 2, 3, 4, \ldots\}$ be the set of integers (positive and negative whole numbers and zero). Then $\mathbb{Z} \times \mathbb{Z}$ will be the set of all ordered pairs (a, b) where both a and b are integers. We could draw a picture of this set by representing one copy of $\mathbb{Z}$ on a horizontal line, the other copy on a vertical line and marking with dots the ordered pairs as follows:-

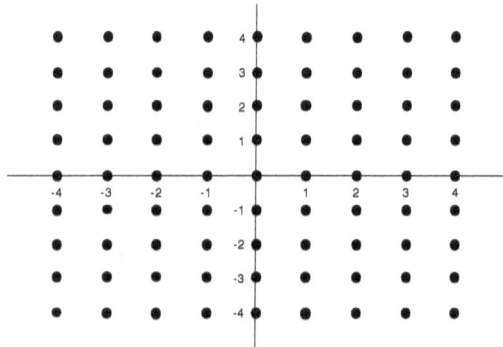

Figure 1.1

Relations

This is the last topic from set theory that we are going to need. Suppose we have a set A, and that we can define some property that 'links' two members of the set with each other, then that property is known as a relation. For example, suppose we let A be the set of all letters in the alphabet and we say that two letters are 'linked' (related) if either both are vowels or both are consonants.

So the letter a is related to the letters e, i, o, u and of course to itself(!) but to none of the other letters, and the letter b is related to each of the letters $c, d, f, g, h, j, k, l, m, n, p, q, r, s, t, v, w, x, y, z$ and itself but to no others.

Now most of the examples of sets we are going to be using will be sets of numbers and so most relations we meet will be relations defined on numbers. Amongst these, some of the most common are:-

a is related to b if $a = b$
a is related to b if $a < b$
a is related to b if $a > b$
a is related to b if $a \neq b$
a is related to b if $a \leq b$
a is related to b if $a \geq b$

But of course we can define a relation any way we want, for example we could say:-

a is related to b if $a - b$ is exactly divisible by 2

Rather than keep on writing out a is related to b , I will use the shorthand form $a\rho b$ in future. (ρ is the Greek letter rho, which is the same as our letter "r")

Now I am going to give the formal definition of a relation in the few lines below (in between the "*'s"). This can be omitted at first reading.

****

Definition 1.3.4. *A **relation** defined on a set A is a subset of the set $A \times A$.*

In other words, a relation is just a set of ordered pairs where both the first and second entries in the pair are drawn from the same set.

CHAPTER 1. SETS, NUMBERS AND FUNCTIONS

****

There is a special class of relations called **transitive relations** that we are going to use later on in the book. So I will define those relations here.

Definition 1.3.5. *A relation ρ, is called **transitive** if whenever $a\rho b$ and $b\rho c$ we can conclude that $a\rho c$.*

Let's look at the seven relations defined above.
The "=" relation is obviously transitive, and since if $a < b$ and $b < c$, we know that $a < c$ so is the "<" relation. The ">" relation is also transitive. However the "$\neq$" relation is not transitive, because we can have $a \neq b$ and $b \neq c$, yet a and c can be equal!
If $a \leq b$ and $b \leq c$ then a will be less than or equal to c so the "$\leq$" relation is transitive. By a similar argument the "$\geq$" relation is transitive as well.
What about the last of the seven? Well if $a - b$ is exactly divisible by 2 and $b - c$ is also exactly divisible by 2, we could write

$$a - b = 2m \text{ and } b - c = 2n, \text{ where } m \text{ and } n \text{ are integers}$$

Hence
$$a - c = (a - b) + (b - c) = 2m + 2n = 2(m + n)$$

will also be exactly divisible by 2. So this relation is also transitive.

(There are other special classes of relations besides transitive, but this is the only one that we need, so I am not going to mention the others!)

Examples 1.3.3
1. Define a relation on the set of numbers, ρ by:- $x\rho y$ if $x^2 + y^2 = 1$.
This relation is not transitive because $1^2 + 0^2 = 1$ and $0^2 + (-1)^2 = 1$ but $1^2 + (-1)^2 \neq 1$
So $1\rho 0$ and $0\rho(-1)$ but 1 is not related to (-1)

2. Define a relation on the set of numbers greater than 1, ρ by:- $x\rho y$ if $x^2 < y$
This relation is transitive because if $x^2 < y$ and $y^2 < z$, then since $y < y^2$, we can conclude that $x^2 < z$.

1.4. Numbers: Whole, Rational and Real

I am assuming everyone thinks they know what a number is, even if actually they don't, so I will be examining numbers through the properties they have rather than attempting any kind of formal definition.

Just to make sure we all understand the words we use, I shall refer to the set
$$\{0, 1, 2, 3, 4, 5, \ldots\}$$
as the Natural Numbers, and for these we will use the symbol $\mathbb{N}$; the set
$$\{\cdots -3, -2, -1, 0, 1, 2, 3, \ldots\}$$
is called the Integers, with symbol $\mathbb{Z}$; and the set
$$\{p/q, \text{ where both } p \text{ and } q \text{ belong to } \mathbb{Z}\}$$
is called the Rational Numbers, written as $\mathbb{Q}$.

Notice that I am not able to give a simple definition of the set of real numbers.

But we are all aware of the simple proof that the square root of 2 is not a rational, it can only be represented as a non-terminating decimal fraction $1.41421356237\ldots$ and it is these non-terminating decimals (which include other famous numbers such as e and π) which make up the real numbers, $\mathbb{R}$.

Lastly the set of numbers of the form
$\{a + bi, \text{ where } i^2 = -1 \text{ and } a \text{ and } b \text{ belong to } \mathbb{R}\}$
are called the complex numbers, $\mathbb{C}$. I shall not be using complex numbers in the rest of the book but will use them in this section as examples only.

In fact, there is a wonderful book entitled "Mathematics made Difficult" which spends several chapters working towards a definition of the natural numbers and then a proof that 1+1=2. Defining these numbers is actually much more difficult than you might think!

The first and probably most important property of numbers is that we can do arithmetic with them.
You might think of arithmetic as addition, subtraction, multiplication and division but these operations don't all make sense for the Natural Numbers or even the Integers. For example you cannot subtract 4 from 2 in the Natural Numbers. It is to enable us to do just this kind

of operation that we have to use the Integers, where we regard subtraction of an integer as the addition of the negative of that integer. Hence:

$$2 - 4 = 2 + (-4) = -2$$

Unfortunately, although we can multiply integers together, we cannot always divide one integer by another. Certainly $8/2 = 4$, but what about 8/3? It is to enable us to do just this kind of operation that we have to use the Rationals, where we regard division by a number as multiplication by the reciprocal of that number. Hence:

$$8/3 = 8 * (1/3) = 2\frac{2}{3}$$

So if we really want to do arithmetic we have to work with the Rational Numbers. Lets try to codify the rules for arithmetic.

Remember we only need to talk about addition and multiplication.

Rules for addition

We can add two numbers together to get a third.
The order in which we add numbers doesn't matter.
There is a special number, called zero, which if we add it to any other number does not change that number.
And for every number we can find another number which if we add the two together gives that special number zero.

In symbols we can say:-

Let $\mathbb{F}$ be a set on which the binary operation $+$ is defined.
If a and b belong to $\mathbb{F}$, then $a + b$ is in $\mathbb{F}$. (1.1)
For all a and b in $\mathbb{F}$, $a + b = b + a$ (1.2)
For all a, b and c in $\mathbb{F}$, $(a + b) + c = a + (b + c)$ (1.3)
There is a number 0, such that $a + 0 = a$ for all a in $\mathbb{F}$ (1.4)
For every a in $\mathbb{F}$, there is another element of $\mathbb{F}$, written $-a$, (1.5)
such that $a + (-a) = 0$ (1.6)

Rules for multiplication
We can multiply two numbers together to get a third.

The order in which we multiply numbers does not matter.
There is a special number, called one, which if we multiply any other number by it does not change that number.
And for every number we can find another number which if we multiply the two together gives that special number one.

In symbols we can say:-

Let $\mathbb{F}$ be a set on which the binary operation $\cdot$ is defined.

If a and b belong to $\mathbb{F}$, then $a.b$ is in $\mathbb{F}$. (1.7)

For all a and b in $\mathbb{F}$, $a.b = b.a$ (1.8)

For all a, b and c in $\mathbb{F}$, $(a.b).c = a.(b.c)$ (1.9)

There is a number 1 such that $a.1 = a$ for all a in $\mathbb{F}$ (1.10)

For every a in $\mathbb{F}$, there is another element of $\mathbb{F}$, written $1/a$, (1.11)

such that $a.(1/a) = 1$ (1.12)

Lastly we can combine addition and multiplication as follows:

Let $\mathbb{F}$ be a set on which the binary operations $+$ and $\cdot$ are defined.

For all a, b and c in $\mathbb{F}$, $a.(b + c) = a.b + a.c$ (1.13)

Given all these axioms, as they are called, we can derive all the properties of arithmetic with which we are familiar. So let us make a formal definition.

Definition 1.4.1. *Let $\mathbb{F}$ be a set on which two binary operations, written $+$ and $\cdot$ are defined. If the two operations obey the axioms (1.1) to (1.11) then $\mathbb{F}$ is called a field.*

Now $\mathbb{Q}$ the set of rational numbers, $\mathbb{R}$ the set of real number and $\mathbb{C}$ the set of complex numbers are all fields. Notice that $\mathbb{N}$ and $\mathbb{Z}$ are not fields.

However, we haven't yet captured all of the subtlety that (some) numbers can have. For integers, rationals and reals, have another property. They can be put into an order, $\ldots 3 < -2 < -1 < 0 < 1 < 2 < 3 < \ldots$ for example. In fact given any two numbers a and b, then $a < b$ or $b < a$ or a and b are the same number.

Again, more formally:

Definition 1.4.2. *If there exists a relation, written $<$, defined on a set S, such that for any a and b belonging to the field, $a < b$ or $b < a$ or $a = b$, then we say that S is an <u>ordered set</u>.*

We need to connect this idea of an ordering, $<$ with the arithmetic operations and this is done as follows:-

Definition 1.4.3. *Let $\mathbb{F}$ be a field with a relation $<$ defined on it so that if a, b and c all belong to $\mathbb{F}$ and the following two properties hold:*
(1) if $a < b$ then $a + c < b + c$
and
(2) if $a < b$ and $0 < c$ then $a.c < b.c$

Then $\mathbb{F}$ *is called an <u>ordered field</u>.*

Now, notice that $\mathbb{C}$, the complex numbers fails this property. There is no obvious definition of "less than" for complex numbers, I mean is i greater than or less than $-i$? So it is not an ordered field. Try and prove this for yourself. But $\mathbb{Q}$ and $\mathbb{R}$ are both ordered fields. A lot of the proofs to follow in later chapters will use the fact that $\mathbb{R}$ is an ordered field and so they would work equally well for the rationals too.

Finally, we come to a property that the real numbers have but which the rational numbers do not.
But I need some more definitions first.

Definition 1.4.4. *Let $\mathbb{F}$ be an ordered field.*
*Then a subset S of $\mathbb{F}$ is **bounded above** if there exists an m in $\mathbb{F}$, such that for every x in S, $x < m$.*
*A subset S of $\mathbb{F}$ is **bounded below** if there exists an n in $\mathbb{F}$, such that for every x in S, $n < x$.*
*A subset S of $\mathbb{F}$ is **bounded** if it is bounded above and below.*

Given a subset S of $\mathbb{F}$ which is bounded above, the **supremum**, or least upper bound, of S is a number k, which
(1) is an upper bound for S
and

(2) no other number that is smaller than k is an upper bound for S.

*Similarly if S is bounded below we can define the **infimum**, or greatest lower bound, of S as a number k, which
(1) is a lower bound for S
and
(2) no other number that is larger than k is a lower bound for S.*

We write sup S for the supremum of S and inf S for the infimum of S.

Examples 1.4.1
1. $(0,1) = \{x$ belongs to $\mathbb{R}$ and $0 < x < 1\}$ is bounded above and below. Its supremum is 1 and its infimum is 0. Notice that neither the supremum or the infimum belong to $(0,1)$.

2. $[0,1] = \{x$ belongs to $\mathbb{R}$ and $0 \leq x \leq 1\}$ is bounded above and below. Its supremum is 1 and its infimum is 0. Notice that both the supremum and the infimum belong to $[0,1]$.

3. $S = \{x$ is in $\mathbb{Q}$ and $x^2 \leq 2\}$ is bounded above and below. Note that as S is regarded as a subset of $\mathbb{Q}$, the supremum of S, the square root of 2, does not exist, in other words S is bounded above but does not have a supremum in $\mathbb{Q}$.

4. $S = \{x$ is in $\mathbb{R}$ and $x^2 \leq 2\}$ is bounded above and below. Now that S is regarded as a subset of $\mathbb{R}$, the supremum of S, the square root of 2, does exist, in other words S is bounded above and does have a supremum in $\mathbb{R}$.

So now we have our property that will distinguish real numbers from rationals. It is called the completeness property and says:-

Definition 1.4.5. *An ordered field is complete if every subset which is bounded above has a supremum, (a least upper bound).*

1.5. Some Properties of the Reals

I shall now prove a few results I am going to need later on. Although the section is called Some Properties of the Reals these will in fact be true for any Complete Ordered Field.

Define the sets $A + B$ and $A.B$ by:

$$A + B = \{a + b \text{ where } a \text{ is in } A \text{ and } B \text{ is in } B\}$$

$$A.B = \{a.b \text{ where } a \text{ is in } A \text{ and } B \text{ is in } B\}$$

Our first result is:

Theorem 1.5.1. *Let A and B be two non-empty subsets of $\mathbb{R}$ which are both bounded above. Then:*
(1) $\sup(A + B) = \sup A + \sup B$
(2) $\sup(A.B) = \sup A . \sup B$

Proof. (1) Let $m = \sup A$ and $n = \sup B$. Now for any x in $A+B$, $x = a_0 + b_0$, where a_0 is in A and b_0 is in B. Hence $a_0 \leq m$ and $b_0 \leq n$. It follows that $a_0 + b_0$ must be less than or equal to $m + n$, so $m + n$ is an upper bound for $A + B$.
Now we have shown that $A + B$ is bounded above, we know there is a least upper bound, call it k.
It follows that $k \leq m + n$.
On the other hand $c \leq k$ for all c in $A + B$ and so $c - n \leq k - n$ for any c in $A + B$.
But any a in A can be written as $a = c - n$, for some c in $A + B$ (to see this just add n to a, you must get a number in $A + B$).
So, because $A = \{c - n \text{ for } c \text{ in } A + B\}$, $k - n$ must be an upper bound for A so $m \leq k - n$.

So we have shown both $k \leq m+n$ and $k \geq m+n$. Hence $k = m+n$.

I will leave the proof of part (2) to you as an exercise.

□

A lot of the discussion that is going to follow in subsequent chapters will be be based on the ideas of an *interval* of the real numbers. So let us just take a moment to define exactly what we mean by an interval and to discuss one or two of the simplest properties that intervals have.

Definition 1.5.1. *An <u>interval</u>, I, is a set of real numbers with the property that if any two numbers x and y belong to I then so does every number between x and y.*

1.5. SOME PROPERTIES OF THE REALS

We use the following notation for the four possible different kinds of bounded intervals (those that have both a maximum and a minimum):-

$$(a,b) = \{x \text{ such that } a < x < b\}$$
$$[a,b] = \{x \text{ such that } a \leq x \leq b\}$$
$$(a,b] = \{x \text{ such that } a < x \leq b\}$$
$$[a,b) = \{x \text{ such that } a \leq x < b\}$$

There are also four kinds of unbounded intervals which we represent by:-

$$(a,\infty) = \{x \text{ such that } x > a\}$$
$$[a,\infty) = \{x \text{ such that } x \geq a\}$$
$$(-\infty,b] = \{x \text{ such that } x \leq b\}$$
$$(-\infty,b) = \{x \text{ such that } x < b\}$$

We call the intervals $(a,b), (a,\infty)$ and $(-\infty,b)$ *open intervals*, and the intervals $[a,b], [a,\infty)$ and $(-\infty,b]$ *closed intervals*. The intervals $(a,b]$ and $[a,b)$ are often called *half-open* (or half-closed).

It is the open and closed bounded intervals that are going to be most important for us in the rest of the book.

When it is important to distinguish open from closed intervals in a diagram, I will use the following notation.

Figure 1.2: Open Interval, (a,b)

Figure 1.3: Closed Interval, $[a,b]$

We will also be talking a lot about neighbourhoods of points in the chapters to come, so a proper definition of that term is also needed.

Definition 1.5.2. *A neighbourhood of a point is any open interval that contains that point.*

You should also be aware that the terms *open* and *closed* are of considerable importance in many other branches of mathematics, for example Topology. It is possible, and quite easy, to give a definition of what we mean by an open set of numbers, not just an interval, in a way that can be generalised to other sets than just the real numbers.

To see how we might do this, think about a closed bounded interval $[a, b]$ for a moment. The points a and b are called the *boundary points* of the interval, for obvious reasons! However a and b are also the boundary points of the open interval (a, b), even though they don't belong to that interval. (See Figures 1.2 and 1.3 above).
So we could say that a closed interval contains its boundary points and an open one does not. Note this is true also of the unbounded closed and open intervals, where there is only one boundary point for such an interval. (Remember there is no such number as ∞, it is just a useful notation).
It doesn't seem that this is very startling, in fact we seem to be arguing around in a circle, but, if we could define a boundary point another way, then we could construct an independent definition of open and closed sets. See Exercise 1.8, questions 7 to 10, where this idea is developed further.

To help us here I would like to remind you of the following elementary idea: the MODULUS or ABSOLUTE VALUE of a real number x, written $|x|$, pronounced 'mod x', is defined as

$$|x| = \begin{cases} x, & \text{if } x \geq 0; \\ -x, & \text{if } x < 0. \end{cases}$$

It has the following important properties:

$$|x| \geq 0 \tag{1.14}$$
$$|x| = 0 \text{ if and only if } x = 0 \tag{1.15}$$
$$|xy| = |x|.|y| \tag{1.16}$$
$$|x + y| \leq |x| + |y| \text{ (called the TRIANGLE INEQUALITY)} \tag{1.17}$$

The first two statements are obvious from the definition of $|x|$.
The third property I leave to you as an exercise.
To prove the triangle inequality notice that

$$-|x| \leq x \leq |x|$$
$$\text{and}$$
$$-|y| \leq y \leq |y|$$

Hence $-(|x|+|y|) < x+y < |x|+|y|$

Then either
$$0 \leq x+y \leq |x|+|y|, \text{ so } |x+y| \leq |x|+|y|$$
or
$$0 \geq x+y \geq -(|x|+|y|)$$
then (multiplying by -1 and reversing all the signs of the inequality)
$$0 \leq -(x+y) \leq -(|x|+|y|)$$
in which case $|x+y| \leq |x|+|y|$ again, as we wanted.

We are coming to some more complex ideas now and firstly I want to be able to talk about two sets, in fact I shall only consider intervals, being *next to each other* or rather *contiguous*.

Definition 1.5.3. *Two sets, A and B, are contiguous if the sets overlap (that is $A \cap B$ is non-empty), or else a boundary point of one set belongs to the other.*

From that definition we can derive a useful property that will help us decide when two intervals are contiguous. Here it is.

Theorem 1.5.2. *If A and B are intervals and if there is a point x_0 in A, not in B, such that any neighbourhood of x_0 intersects B, then x_0 must be a boundary point for B and hence A and B are contiguous.*

Proof. If B is bounded, let the boundary points of B be p and q with $p < q$.
If B is unbounded, let p be its one and only boundary point.
As x_0 does not belong to B, then either x_0 is greater than every x in B or x_0 is less than every x in B.

CHAPTER 1. SETS, NUMBERS AND FUNCTIONS

So we will assume $x_0 < x$ for all x belonging to B. (The proof for the other case is exactly the same).

Then, if $p \neq x_0$ we have that $p - x_0 > 0$, so $|p - x_0| = \varepsilon$, say.

Figure 1.4

So the set $\{x,\text{ such that }|x - x_0| < \varepsilon/2\}$ will be a neighbourhood of x_0 which cannot intersect B. However this contradicts the condition given in the theorem. Hence $p = x_0$, in other words x_0 is a boundary point of B. It follows from the definition that A and B must be contiguous.

□

The next theorem seems a little strange, not just because its name sounds like something from a horror movie rather than a mathematics text! We shall be using it a couple of times in the next chapter to help us prove some important results. Although you should read the proof carefully and try to understand it, it is not *essential* that you do that here and now. You can take the result on trust if you like, see how it is used in Chapter 2 and then return later to work through its proof. Up to you.

Before stating it formally, let me try to explain exactly what it is saying.

Suppose we have a closed interval with a transitive relation defined on it in such a way that if we take a neighbourhood of x, (any point in the interval, then every point in that neighbourhood (that is also in the interval), will be related to x. So if we take x as the end point a, then take a neighbourhood of a. Any point in that neighbourhood which is also in the interval, will be related to a. Then if we were to take a point in that neighbourhood, c say, and take a neighbourhood of c then, this time every point in this new neighbourhood is related to c. We now creep a little further along and take a point $d > c$ in our neighbourhood of c, then take a neighbourhood of d and if we find that every point in that neighbourhood is related to d.......... We just continue this process creeping along the original interval by these small neighbourhoods and eventually we will reach b, the other end of the interval, and we can conclude, since the relation is transitive, that a is related to b. By doing this, we can turn a "local" relation into a "global" one. Here comes the formal statement.

Theorem 1.5.3. *The Creeping Lemma*
If we have a relation ρ defined on a closed interval $[a, b]$ such that:
(1) ρ is transitive
and
(2) every x in $[a, b]$ has a neighbourhood, N_x, such that whenever u and v belong to $N_x \cap [a, b]$ we know that $u\rho v$
then we can conclude that $a\rho b$

Proof. Let $M = \{x \text{ such that } a\rho x\}$. If we let $x = u = v = a$ in condition (2) of the theorem we get that $a\rho a$, so M is not empty.

So M is a non-empty subset of $[a, b]$, hence it is bounded above by b. This means that it will have a least upper bound, which we will call m.

Now choose $\varepsilon > 0$ and let N_m be the neighbourhood of m given by $(m - \varepsilon, m + \varepsilon)$.

Since m is the supremum of M, it follows that $m - \varepsilon$ is not an upper bound for M, so there must be some number u in M such that $u > m - \varepsilon$.

The number u has the properties that $a\rho u$, since u is in M and u belongs to $N_m \cap [a, b]$.

I am now going to prove that b must belong to N_m.
Suppose that b is not in N_m, see Figure 1.5 below. Then let $v = m + \frac{1}{2}\varepsilon$, so v belongs to $N_m \cap [a, b]$.

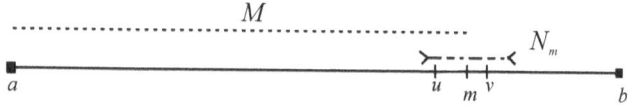

Figure 1.5

This means, using condition (2) of the theorem, that $u\rho v$.
We already know that $a\rho u$ and we also have that ρ is transitive, hence $a\rho v$. This means that v must belong to M, but this is impossible

because $v > m$ and m is the supremum of M.
So our supposition must be wrong and b must belong to N_m.
Using condition (2) once more, noting that u belongs to $N_m \cap [a,b]$ and b belongs to N_m, we deduce that $u\rho b$.
So as $a\rho u, u\rho b$ and ρ is transitive, we conclude that $a\rho b$ as required. □

1.6. Definition and Simple Properties of Functions

Analysis is concerned with functions. Whether it is proving they are continuous, differentiable or integrable, functions are central. So it makes sense to say exactly what we mean by a function.

A **function** f from a set A to another set B defines a rule which assigns to each x in A a unique element y in B. We write $f: A \longrightarrow B$ to represent the function and $f(x) = y$ to show the assignment of an individual element.
The element y is known as the *image* of x.

Please note that this is not a formal definition of a function. And also please note I have not said that a function is a rule merely that it defines one! For what follows in this book, we do not need that formal definition, but just in case you are curious here it is anyway:-

Definition 1.6.1. *A function f from a set A to another, possibly the same, set B is a set of ordered pairs (x, y), no two of which can have the same first element and where x belongs to A and y belongs to B. In other words this is just a subset of $A \times B$.*

You should be able to see how this formal definition ties up with the more friendly explanation given previously.

When A and B are subsets of the real numbers we can draw the graph of the function by plotting those ordered pairs (x, y).
Note that the condition given that each x in A is assigned to a unique y in B means that any vertical line drawn through a point of A will meet the graph at most once.

1.6. DEFINITION AND SIMPLE PROPERTIES OF FUNCTIONS

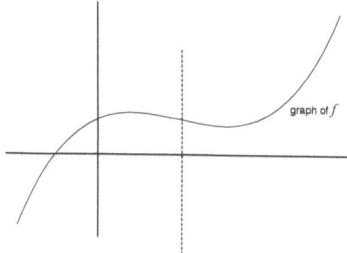

Figure 1.6

Now for some more important ideas we will need.

The **domain** of a function is the largest set on which that function is defined.

If you think about the graph of a function, the domain is the set of points on the x-axis that lie vertically above or below points of the graph.

The **image** of a set, S, under a function f, is the set of values of $f(x)$ for all x belonging to S.

The **range** of a function is the image of its domain.

So if $f : A \longrightarrow B$ is a function, then its domain is A and its range is $f(A)$ (which must be a subset of B).

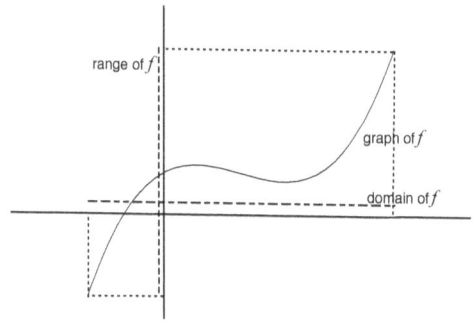

Figure 1.7

25

Consider a subset T of the range B of a function, f. The set of values in A that are assigned to T by the function is called the **inverse image** of T.

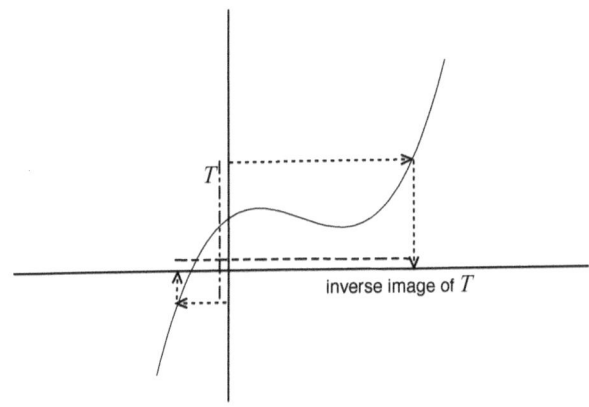

Figure 1.8

Lastly a **polynomial function** is a function from $\mathbb{R}$ to itself, given by the assignment

$$f(x) = a_0 + a_1 x + a_2 x^2 + a_3 x^3 + \cdots + a_n x^n$$

Let us look at some examples.

Examples 1.6.1
1. $f(x) = x^2$ defines a polynomial function from $\mathbb{R}$ to itself.

1.6. DEFINITION AND SIMPLE PROPERTIES OF FUNCTIONS

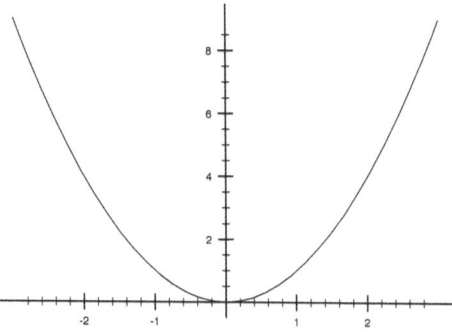

Figure 1.9

The domain of f is $\mathbb{R}$ and its range is the set of non-negative real numbers.

2. $f(x) = \sqrt{x}$ may or may not define a function depending on how you interpret $\sqrt{x}$. After all both 2 and -2 are the square root of 4, so if we allow negative square roots we would get a graph as follows.

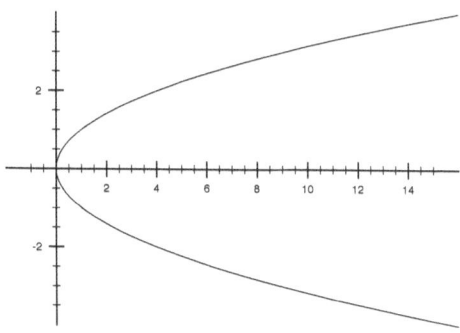

Figure 1.10

You can see immediately that this not the graph of a function. If we only allow positive square roots, however, we get the following graph, which is that of a function with domain the set of non-negative reals and range the same set.

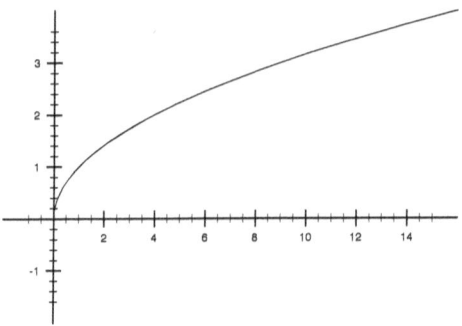

Figure 1.11

3. $f(x) = x^3 - 3x^2 + 2x$ is a function from $\mathbb{R}$ to itself. It is a polynomial function. Both its domain and range are the set $\mathbb{R}$.

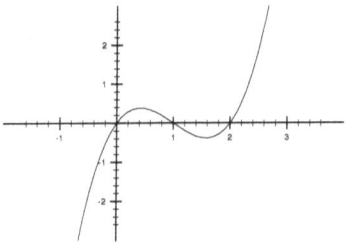

Figure 1.12

4.
$$f(x) = \frac{x^2 + 1}{x^2 - 1}$$

defined on the set of real numbers excluding 1 and -1 (we often write $\mathbb{R} \setminus \{1, -1\}$ for this) is a function, known as a **rational function** because it is made up of one polynomial function divided by another.

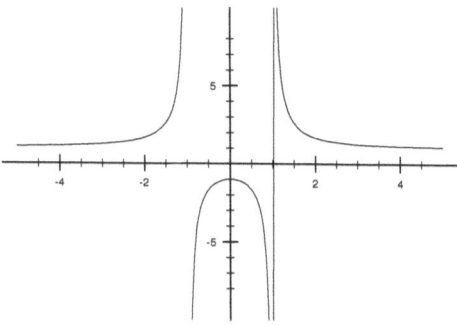

Figure 1.13

The domain of f is $\mathbb{R} \setminus \{1, -1\}$, and its range is the set consisting of all positive reals greater than 1 and all negative reals less than or equal to -1.

1.7. Further Properties of Functions

The inverse of a function

Sometimes we want to be able to "go the other way", that is go back from B to A. In some circumstances this is possible. We have already seen what the "inverse image of a set" means and what we would like to say is that if $f(x) = y$ then we will define the inverse function f^{-1} by $x = f^{-1}(y)$. However this is only possible if for each value of x the original function f gives at most one value for $f(x)$. (We say such a function is one to one, written 1-1).

A function $f : A \longrightarrow B$ is called **one to one** if for each value of x in A there is at most one value y in B such that $f(x) = y$.

Geometrically, a function f is 1-1 if both every horizontal line and every vertical line meets the graph of f at most once. (Why?)

A 1-1 function f has an **inverse function**, f^{-1} defined by $x = f^{-1}(y)$ whenever $f(x) = y$

Examples 1.7.1

1.
$$f(x) = x$$

This is a 1-1 function, if you draw the graph, you will see this clearly but we really ought to prove it.
Given any x in $\mathbb{R}$, as $f(x) = x$, there is only one possible value for $f(x)$.
Hence f is 1-1 and has an inverse function.
The inverse function f^{-1} is defined by $f^{-1}(y) = y$, in other words f^{-1} is the same function as f! Note that $f(x) = x$ is is the only function for which the function and its inverse are the same.

2.
$$f(x) = x^2$$

This function is not 1-1 because, for example, $f(-2) = f(2) = 4$.
Hence it does not have an inverse function. (Compare with Examples 1.6.1, number 2 above.)

3.
$$f(x) = x^3$$

This is a 1-1 function, because suppose there was some two numbers a and b such that $f(a) = f(b)$. This would mean that $a^3 = b^3$, in other words $a^3 - b^3 = 0$.
Now let's do some simple algebra:-

$$(a^3 - b^3) = (a - b)(a^2 + ab + b^2)$$

and since $a^2 + ab + b^2$ is only equal to zero if both $a = 0$ and $b = 0$ (prove this), the fact that $a^3 - b^3$ is zero means the only possibility is that $a - b = 0$, that is $a = b$
(see Figure 1.11 for the graph of f).

4.
$$f(x) = \frac{x-1}{x+1} \text{ for } x \geq 0$$

1.7. FURTHER PROPERTIES OF FUNCTIONS

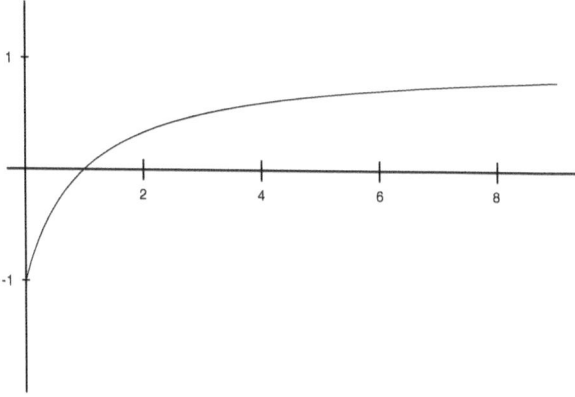

Figure 1.14

From the graph, you can see that f is 1-1 and therefore will have an inverse. However once again, we need to prove this. Suppose $f(a) = f(b)$. In other words:

$$\frac{a-1}{a+1} = \frac{b-1}{b+1}$$

Then, noting that since both a and b are greater than or equal to zero, $a+1$ and $b+1$ are both positive,

$$(a-1)(b+1) = (b-1)(a+1)$$

Multiply out the brackets to get

$$ab - b + a - 1 = ab - a + b - 1$$

So

$$a - b = -(a - b)$$

But the only number which is equal to its own negative is zero, so $a - b = 0$ and thus $a = b$ as required.
Now we need to find the inverse function.
Suppose

$$y = \frac{x-1}{x+1}$$

CHAPTER 1. SETS, NUMBERS AND FUNCTIONS

Then, rewriting this equation and solving it for x we get:

$$xy + y = x - 1$$
$$x - xy = y + 1$$
$$x(1 - y) = 1 + y$$
$$x = \frac{1+y}{1-y}$$

Hence $f^{-1}(y) = \frac{1+y}{1-y}$ is the inverse function.

The sum, product and quotient of two functions

I have already been using the ideas of the sum, the product and even the quotient of two functions in the examples at which we have been looking. Now I shall formally define them, in the obvious way.

Definition 1.7.1. *If $f : S \longrightarrow \mathbb{R}$ and $g : T \longrightarrow \mathbb{R}$ are functions then we construct new functions:*

$$f + g : S \cap T \longrightarrow \mathbb{R} \text{ defined by } (f+g)(x) = f(x) + g(x)$$

and

$$f.g : S \cap T \longrightarrow \mathbb{R} \text{ defined by } (f.g)(x) = f(x).g(x)$$

and

$$(f/g) : S \cap T' \longrightarrow \mathbb{R} \text{ defined by } (f/g)(x) = f(x)/g(x)$$

where $T' = T \cap \{x \text{ for which } g(x) \neq 0\}$

If you look at the examples above you will see that we have met all of these constructions already, for instance Examples 1.6.1, number 4 and 1.6.2, number 4 are both examples of the quotient of two functions.

The composition of two functions

Lastly, we are going to define the composition of two functions. This is a formal way of saying we perform one function then perform another on the results of the first. The definition is:

Definition 1.7.2. *Let $f : S \longrightarrow \mathbb{R}$ and $g : T \longrightarrow \mathbb{R}$ be functions such that $g(x)$ belongs to S for every x in T, then we can define a new function:*

$$f \circ g : T \longrightarrow \mathbb{R} \text{ defined by } (f \circ g)(x) = f(g(x))$$

*called the **composition** of f and g.*

Notice that, in general $f \circ g \neq g \circ f$.

Examples 1.7.2
Let
$$g(x) = x^2 \text{ and } f(x) = \frac{1-x}{1+x^2}$$
both functions defined for all values of x in $\mathbb{R}$.

1. Then $f \circ g : \mathbb{R} \longrightarrow \mathbb{R}$ will be given by:
$$(f \circ g)(x) = \frac{1-x^2}{1+(x^2)^2} = \frac{1-x^2}{1+x^4}$$

2. Note that $g \circ f : \mathbb{R} \longrightarrow \mathbb{R}$ will be given by:
$$(g \circ f)(x) = \left(\frac{1-x}{1+x^2}\right)^2 = \frac{1-2x+x^2}{1+2x^2+x^4}$$

1.8. Exercises

Sets and Relations
1. Let $A = \{1, 3, 5, 7, 9\}$ and $B = \{2, 4, 6, 8\}$. Then write down all the elements of $A \times B$. (You should have 5 x 4 = 20 distinct elements).

2. Let $\mathbb{N}$ be the set of natural numbers, that is $\{0, 1, 2, 3, 4, 5 \ldots\}$. Define a relation ρ on $\mathbb{N}$ by $a\rho b$ if and only if $a - b$ is even. (Notice that it follows that if $a\rho b$ then $a \geq b$). Either prove that ρ is transitive or give an example to show why it is not.

3. Let $\mathbb{N}$ be the set of natural numbers, that is $\{0, 1, 2, 3, 4, 5 \ldots\}$. Define a relation ρ on $\mathbb{N}$ by $a\rho b$ if and only if $a - b$ is odd. (Notice that it again follows that if $a\rho b$ then $a \geq b$). Either prove that ρ is transitive or give an example to show why it is not.

Numbers

4. Prove that $\mathbb{C}$, the complex numbers, is not an ordered field. (Hint: Assume there is an ordering which obeys the rules set out in Definition 1.4.3, then show that either of the statements $i < 0$ or $i > 0$ leads to a contradiction.)

5. Find the supremum and infimum, where they exist, of the following sets of real numbers:
(a) The set of numbers whose cube is less than or equal to 8.
(b) The set of numbers x, where $x = 1/(1+t^2)$, for t any real number.
(c) The set of numbers x, where $x = t/(1+t^2)$, for t any real number.

6. Prove that if A and B are two non-empty subsets of $\mathbb{R}$ which are both bounded above, then $\sup(A.B) = \sup A. \sup B$.

The following four exercises generalise the idea of open and closed sets given in the text.

7. Define the distance between any two points, x and y, that both belong to $\mathbb{R}$ by $|x - y|$. We write this as $d(x, y)$. Show that:
(a) $d(x, y) \geq 0$
(b) $d(x, y) = 0$ if and only if $x = y$
(c) $d(x, y) = d(y, x)$
(d) $d(x, z) \leq d(x, y) + d(y, z)$

8. Define the distance between a real number x and a set of real numbers A, written $d(x, A)$, by:-

$$d(x, A) = \inf\{d(x, y) \text{ for all values of } y \text{ in } A\}$$

Show that:
(a) $d(3, A) = 2$, where $A = [0, 1]$
(b) $d(1, B) = 0$, where $B = (0, 1)$
(c) $d(0, C) = 0$, where $C = \{1/n \text{ for all values of } n \text{ in } \mathbb{N}\}$

9. Define a *boundary point* of a set A as a point x where both $d(x, A) = 0$ and $d(x, \mathbb{R} \setminus A) = 0$.
Show that:

(a) 0 is a boundary point for the set $(0, 1)$
(b) 0 is a boundary point for the set $[0, 1]$
(c) The set of boundary points of C where $C = \{1/n \text{ for all values of } n \text{ in } \mathbb{N}\}$ is $C \cup \{0\}$.

10. Define a set A to be *open* if it contains none of its boundary points. Define a set A to be *closed* if it contains all of its boundary points. Show that:
(a) $(0, 1)$ is open.
(b) $[0, 1]$ is closed.
(c) The set C, where $C = \{1/n \text{ for all values of } n \text{ in } \mathbb{N}\}$ is neither open nor closed.

Notice that all of the above work relies on the definition of the *distance* between two numbers. So, if we can define the distance between two elements of any set in such a way that the distance obeys the results in Question 7 (a) to (d), we can define open and closed sets by just following through the definitions above.

11. (Harder) Define the distance between two complex numbers $a + bi$ and $c + di$ by

$$d(a + bi, c + di) = \sqrt{|(a^2 + b^2) - (c^2 + d^2)|}$$

Show that this definition satisfies:
a) $d(x, y) \geq 0$
(b) $d(x, y) = 0$ if and only if $x = y$
(c) $d(x, y) = d(y, x)$
(d) $d(x, z) \leq d(x, y) + d(y, z)$
where x, y and z are all complex numbers.
Hence by making the appropriate definitions as suggested above for the distance between a complex number and a set, boundary points, and open and closed sets of complex numbers, show that the set of all complex numbers whose distance is less than 1 from 0 is an open set.

Intervals, Neighbourhoods and Bounds

12. Consider the set $A.B$ given by:
$$A.B = \{a.b \text{ where } a \text{ is in } A \text{ and } B \text{ is in } B\}$$
where A and B are two non-empty subsets of $\mathbb{R}$ which are both bounded above. Prove that:- $\sup(A.B) = \sup A . \sup B$

13. Prove that $|xy| = |x|.|y|$

14. Prove that the sets $A = (0, 1)$ and $B = [1, 2]$ are contiguous.

15. Prove that the sets $A = (0, 1)$ and $B = (1, 2)$ are not contiguous.

16. Suppose that A, a subset of $\mathbb{R}$, is bounded above and that B is a subset of A. Prove that $\sup B \leq \sup A$.

17. Suppose that A, a subset of $\mathbb{R}$, is bounded above. Prove that
$$\inf(-A) = -\sup A$$
where $-A$ is defined by $-A = \{-a; \text{for all values of } a \text{ in } A\}$.

Functions

18. Find the domains of each of the following functions:-

(a) $f(x) = \dfrac{1}{x+1}$

(b) $f(x) = \dfrac{x}{(x-1)(x-2)}$

(c) $f(x) = \dfrac{x-1}{x^2-1}$

Note (a) and (c) are not the same!

19. Find the range and domain of each of the following functions:-

(a) $f(x) = \dfrac{1}{x^2+1}$

(b) $f(x) = \dfrac{x^3}{x^2+1}$

(c) $f(x) = \dfrac{\sqrt{x^4-x^2}}{x}$

20. Find the images of each of the following sets under the functions given in question 19:-

(a) $(-1, 1)$
(b) $[0, \infty)$
(c) $\{0, 1, 2, 3\}$

21. For each of the following functions, f, and sets, X, find $f^{-1}(X)$:

(a) $f(x) = x^3$, $X = (-1.1)$
(b) $f(x) = \dfrac{1}{x+1}$, $X = [0, 2]$
(c) $f(x) = \dfrac{1}{x+1}$, $X = [-2, 2]$

22. For each pair of functions, f and g, given below, find the functions $f.g$, $f \circ g$ and $g \circ f$. In each case give the domain of the function you have found.

(a) $f(x) = x^2$, $g(x) = x + 1$, f and g both have domain $\mathbb{R}$
(b) $f(x) = \dfrac{1-x}{1+x}$, $g(x) = x(1-x)$, f and g both have domain $[0, 1]$
(c) $f(x) = \sqrt{x}$, $g(x) = -\sqrt{x}$, f and g both have domain $[0, \infty)$

23. For the two functions given in question 22 part (b) above, show that f^{-1} exists but g^{-1} does not exist.

Chapter 2
Continuity

2.1. Objectives

In this chapter we meet the concept of "continuity" of a function. The definition given stems from an intuitive idea of what we think continuity may mean to provide a rigorous statement that we can use, involving neighbourhoods. By the end of this chapter you should be able to prove that particular functions are, or are not, continuous, both at particular points and on specified sets. You should understand the make up of continuity proofs and how they relate back to the definition.

You will also have met and understood the ideas of a function being bounded and attaining its bounds together with some of the implications of these ideas.

2.2. New terms and symbols introduced

> a function is continuous on a set
> a function is continuous at a point
> a function is bounded above
> a function is bounded below
> a function is bounded
> a function attains its supremum
> a function attains its infimum

2.3. What does Continuous mean?

Continuity of functions is one of those things that most students feel they understand until it is explained to them and then it suddenly becomes very unclear!

Let me illustrate what I mean. An easy to understand "definition" of a continuous function is that we can draw the graph of the function without taking our pencil off the paper. The graph is "all joined up" so to speak.

This type of "definition" suffers from several problems. First of all, it is almost impossible to express in rigorous mathematical terms, for instance, how are we going to say rigorously, "without taking our pencil off the paper"? Secondly, it only applies to functions from the reals to themselves. I know we are restricting ourselves to such functions in this book, but Analysis as a whole looks at functions from and to a variety of sets and it would be nice to have a definition that generalizes to other cases.

However let us try and use the graph drawing idea to work out what might be a better definition.
Imagine you are drawing the graph of a function defined on the whole of the real numbers. Let us assume that you start on the left hand side of the picture, large negative reals, drawing the graph until you reach the right hand side, large positive reals, then stop.
If you move very slowly from left to right the values of the argument of the function (the numbers being input to the function) won't change very much and, think about this, the values of the images of those arguments cannot change that much either.
Okay, we are beginning to get somewhere. Provided we can explain exactly what we mean by not changing too much, we have the making of a working definition.

What we are saying here is that points, and sets, that start off close to each other should finish up close to each other. Actually we already have a concept of closeness from Chapter 1, where we introduced the idea of two sets being contiguous or "next to each other". So we use that idea to make the following definition:

Definition 2.3.1. *Let S be a subset of $\mathbb{R}$. A function $f : S \longrightarrow \mathbb{R}$ is continuous on S if and only if, for any two subsets A and B of S that are contiguous, $f(A)$ and $f(B)$ are contiguous.*

Now remember that two subsets A and B being contiguous means that either they overlap, in which case there will be an open interval that lies in both A and B, or a boundary point of one set belongs to the other. Using this, we can recast our definition in a way that will actually prove easier to work with!
In fact, it is much easier if we can define continuity at a point in S rather than on the whole of S at once.

Suppose we have a continuous function $f : S \longrightarrow \mathbb{R}$, so f satisfies the property that for any two subsets A and B of S that are contiguous,

$f(A)$ and $f(B)$ are contiguous.
Now consider a point x_0 in S and choose any neighbourhood of $f(x_0)$, $N_{f(x_0)}$.
Clearly x_0 will lie in $f^{-1}\left(N_{f(x_0)}\right)$ and this means that we can choose any neighbourhood of x_0, N_{x_0} and then N_{x_0} and $f^{-1}\left(N_{f(x_0)}\right)$ will be contiguous.
Hence by the continuity of f, $f(N_{x_0})$ and $\left(N_{f(x_0)}\right)$ will be contiguous. See the diagram below for the two possibilities here, (either $f(x_0)$ is an interior point or a boundary point of the set $f(N_{x_0})$).

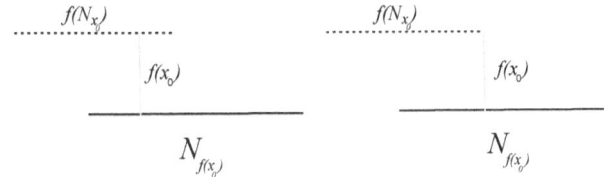

Figure 2.1

In either case, there could be points in $f(N_{x_0})$ not in $\left(N_{f(x_0)}\right)$, but if we shrink N_{x_0} sufficiently all such points must eventually disappear. (Why?) Hence we can construct a neighbourhood of x_0, N_{x_0} such that for each x in $N_{x_0} \cap S$, $f(x)$ is in $N_{f(x_0)}$. In other words if x is not too far from x_0, $f(x)$ will not be too far away from $f(x_0)$.

This is actually a more useful formulation of the definition and will be the one we are going to use. So our final definition of a continuous function will be:

Definition 2.3.2. *The function $f : S \longrightarrow \mathbb{R}$ is continuous at x_0, if, given any neighbourhood $N_{f(x_0)}$ of $f(x_o)$, then there is some neighbourhood N_{x_0} of x_0 such that for each x in $N_{x_0} \cap S$, $f(x)$ is in $N_{f(x_0)}$*

The function $f : S \longrightarrow \mathbb{R}$ is continuous on S if it is continuous at every point x in S.

However although I have shown that the contiguousness definition implies the neighbourhood definition I have not yet shown that the reverse is true.
Actually I do not need to do this since I am going to take Definition 2.3.2 as my *definition* of a continuous function. However, for the sake

of completeness, I am including the argument below. You may skip this section completely if you wish!

****

Optional proof that the neighbourhood condition implies the contiguousness condition for continuity.
Assume the neighbourhood property holds, let S be a subset of $\mathbb{R}$ and let $f : S \longrightarrow \mathbb{R}$ be a function.
Now suppose that we have two subsets A and B of S that are contiguous.
Then either there is an open set (interval) contained in $A \cap B$, call this Case 1,
or,
there is some point x_1 in A, not actually in B but which is on the boundary of B, (i.e an end point for B), Case 2.

Case 1.
Let x_0 belong to $A \cap B$. Then $f(x_0)$ will belong to both $f(A)$ and $f(B)$, hence it will belong to $f(A) \cap f(B)$
Consider $N_{f(x_0)}$, a neighbourhood of $f(x_0)$. From the Neighbourhood Property we know that there is a neighbourhood of x_0, N_{x_0} such that $f(N_{x_0} \cap S)$ lies in $N_{f(x_0)}$. By shrinking the neighbourhood $N_{f(x_0)}$ if necessary, we can choose N_{x_0} to lie in $A \cap B$.
Then for every x in $N_{x_0} \cap S$, $f(x)$ will belong to both $f(A)$ and $f(B)$. So $f(A) \cap f(B)$ contains an open set, namely $N_{f(x_0)}$, hence $f(A)$ and $f(B)$ are contiguous.

Case 2
$f(x_1)$ will belong to $f(A)$. Consider $N_{f(x_1)}$, a neighbourhood of $f(x_1)$. From the Neighbourhood Property we know that there is a neighbourhood of x_1, N_{x_1} such that $f(N_{x_1} \cap S)$ lies in $N_{f(x_1)}$.
Now any neighbourhood of x_1 must intersect B, as x_1 is a boundary (end) point for B. So choose a y in $N_{x_1} \cap S$ such that y is in B.
Then $f(y)$ lies in $f(B)$, but $f(y)$ also lies in $N_{f(x_1)}$. Hence $N_{f(x_1)}$ intersects $f(B)$. This means that any neighbourhood of $f(x_1)$ must intersect $f(B)$, so $f(x_1)$ must be a boundary point for $f(B)$ by definition.

So $f(A)$ and $f(B)$ are contiguous.

End of proof
****

Finally, to make life easier for myself in later sections, I am going to give a number of alternative formulations of the definition of continuity at a point in the following theorem.

Theorem 2.3.1. *The function $f : S \longrightarrow \mathbb{R}$ is continuous at x_0 if any of the following hold:-*

(i) given an ε-neighbourhood N_{f_0}, of $f(x_0)$, then there is some δ-neighbourhood N_{x_0} of x_0 such that, for each x in $N_{x_0} \cap S$, $f(x)$ is in $N_{f_{x_0}}$.

(ii) given any $\varepsilon > 0$, there is some $\delta > 0$ such that if x is in $(x_0 - \delta, x_0 + \delta) \cap S$, then $f(x)$ is in $(f(x_0) - \varepsilon, f(x_0) + \varepsilon)$.

(iii) given any $\varepsilon > 0$, there is some $\delta > 0$ such that if x is in S and $|x - x_0| < \delta$, then $|f(x) - f(x_0)| < \varepsilon$.

Proof. (i) follows from the definition 2.3.2 by noting that any neighbourhood of x_0 will contain a neighbourhood of x_0 that is centred on x_0. Then all we are doing is specifying the size of the neighbourhoods.
(ii) follows from (i) by replacing the ε-neighbourhood and δ-neighbourhood by explicit open intervals.
(iii) follows from (ii) by rewriting the open intervals as inequalities using the modulus or absolute value notation.

$\square$

2.4. Properties of Continuous Functions

In this section we are going to prove some basic results about continuous functions. By the time we get to the end of the section you should have an appreciation of how such proofs are constructed!

We start with singling out two easy functions which are continuous as examples.

Examples 2.4.1
1. The Constant Function is Continuous

The constant function is the function $f : \mathbb{R} \longrightarrow \mathbb{R}$ such that $f(x) = c$ for every x in $\mathbb{R}$.

To prove it is continuous take x_0 in $\mathbb{R}$, then $f(x_0) = c$; so take any neighbourhood of c, N_c. Now we can take *any* neighbourhood of x_0 at all, and for every value of x in that neighbourhood, $f(x)$ will be in N_c, because $f(x) = c$.

Hence the constant function satisfies Definition 2.3.2 and is continuous for any x_0 in $\mathbb{R}$, hence it is continuous.

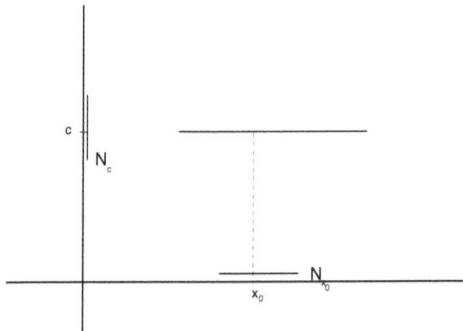

Figure 2.2

2. The Identity Function is Continuous

The identity function is the function $f : \mathbb{R} \longrightarrow \mathbb{R}$ such that $f(x) = x$ for every x in $\mathbb{R}$.

To prove this is continuous take x_0 in $\mathbb{R}$, and any neighbourhood of $f(x_0)$, $N_{f(x_0)}$.

Since $f(x_0) = x_0$, our neighbourhood, $N_{f(x_0)}$ is actually a neighbourhood of x_0!

So let $N_{x_0} = N_{f(x_0)}$, then for every x in N_{x_0}, $f(x)$ will belong to $N_{f(x_0)}$, hence the identity function satisfies Definition 2.3.2 and is continuous for any x_0 in $\mathbb{R}$, hence it is continuous.

2.4. PROPERTIES OF CONTINUOUS FUNCTIONS

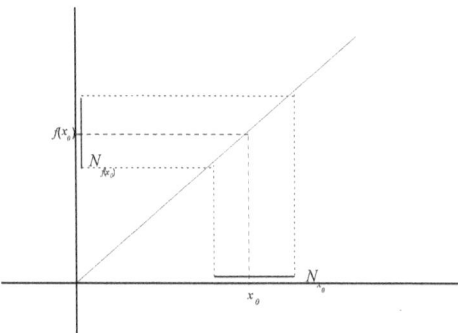

Figure 2.3

Now we come to eight rules that enable us to construct new continuous functions from old.

Theorem 2.4.1. *The Sum of Two Continuous Functions is Continuous*
If $f : S \longrightarrow \mathbb{R}$ and $g : T \longrightarrow \mathbb{R}$ are two continuous functions, then $f + g : S \cap T \longrightarrow \mathbb{R}$ is continuous.

Proof. We prove that if both f and g are continuous at x_0, then $f + g$ is continuous there also. Then the theorem follows easily from that. (Left as an exercise for you).

To prove continuity at x_0 we use Theorem 2.3.1 (i).
Consider an ε neighbourhood of $(f+g)(x_0)$, write this as $N\left((f + g)(x_0), \varepsilon\right)$, then $N\left(f(x_0), \frac{1}{2}\varepsilon\right)$ and $N\left(g(x_0), \frac{1}{2}\varepsilon\right)$ will be neighbourhoods of $f(x_0)$ and $g(x_0)$ respectively.
Hence $N\left(f(x_0), \frac{1}{2}\varepsilon\right) + N\left(g(x_0), \frac{1}{2}\varepsilon\right)$ will be contained in $N\left((f + g)(x_0), \varepsilon\right)$. (Prove this from the definition of $A + B$ in Chapter 1.)

Now because f is continuous at x_0, there is some number δ_1 such that for every x in $N(x_0, \delta_1) \cap S$, $f(x)$ will be in $N\left(f(x_0), \frac{1}{2}\varepsilon\right)$.
Similarly, because g is continuous at x_0, there is some number δ_2 such that for every x in $N(x_0, \delta_2) \cap T$, $g(x)$ will be in $N\left(g(x_0), \frac{1}{2}\varepsilon\right)$.

Now let $N_{x_0} = N(x_0, \delta_1) \cap N(x_0, \delta_2)$.
Then N_{x_0} is a δ-neighbourhood of x_0 such that if x is in $N_{x_0} \cap (S \cap T)$ then $f(x)$ must be in $N\left(f(x_0), \frac{1}{2}\varepsilon\right)$ and $g(x)$ must be in $N\left(g(x_0), \frac{1}{2}\varepsilon\right)$.

This means that $(f+g)(x)$ will be in $N\left((f+g)(x_0),\varepsilon\right)$ as required. Hence $f+g$ is continuous at x_0.

$\square$

Proofs like the one above can appear very daunting when you first meet them, but actually they are structured in a fairly simple and logical manner. To help you see this, I am going to give below the definition of continuity we are using and the the proof again, using different fonts so you can see which part of the definition matches which part of the theorem.

Definition
Given an ε-neighbourhood N_{f_0}, of $f(x_0)$,

then
there is some
δ-neighbourhood
N_{x_0} of x_0

such that,
for each x in $N_{x_0} \cap S$,

$f(x)$ is in $N_{f_{x_0}}$.

Proof
Consider an ε-neighbourhood of $(f+g)(x_0)$,
write this as $N\left((f+g)(x_0),\varepsilon\right)$,
then $N\left(f(x_0),\tfrac{1}{2}\varepsilon\right)$ and $N\left(g(x_0),\tfrac{1}{2}\varepsilon\right)$
will be neighbourhoods of $f(x_0)$
and $g(x_0)$ respectively.
Hence $N\left(f(x_0),\tfrac{1}{2}\varepsilon\right) + N\left(g(x_0),\tfrac{1}{2}\varepsilon\right)$
will be contained in $N\left((f+g)(x_0),\varepsilon\right)$.
Now because f is continuous at x_0,
there is some number δ_1
such that for every x in $N(x_0,\delta_1) \cap S$,
$f(x)$ will be in $N\left(f(x_0),\tfrac{1}{2}\varepsilon\right)$.
Similarly, because g is continuous at x_0,
there is some number δ_2
such that for every x in $N(x_0,\delta_2) \cap T$,
$g(x)$ will be in $N\left(g(x_0),\tfrac{1}{2}\varepsilon\right)$.
Now let $N_{x_0} = N(x_0,\delta_1) \cap N(x_0,\delta_2)$.
Then
N_{x_0} is a δ-neighbourhood of x_0
such that
if x is in $N_{x_0} \cap (S \cap T)$
then $f(x)$ must be in $N\left(f(x_0),\tfrac{1}{2}\varepsilon\right)$ and $g(x)$ must be in $N\left(g(x_0),\tfrac{1}{2}\varepsilon\right)$.
This means that
$(f+g)(x)$ will be in $N\left((f+g)(x_0),\varepsilon\right)$

Hopefully you can now see the structure of the proof. All the following theorems will be proved the same way, it is up to you to find

2.4. PROPERTIES OF CONTINUOUS FUNCTIONS

the relevant parts of the definition in each proof!

Theorem 2.4.2. The Product of Two Continuous Functions is Continuous
If $f : S \longrightarrow \mathbb{R}$ and $g : T \longrightarrow \mathbb{R}$ are two continuous functions, then $f.g : S \cap T \longrightarrow \mathbb{R}$ is continuous.

As you might expect, the proof of this theorem is very similar to that of the previous one.

Proof. We prove that if both f and g are continuous at x_0, then $f.g$ is continuous there also. Then the theorem follows easily from that. (Once again, left as an exercise for you).

To prove continuity at x_0 we use Theorem 2.3.1 (i).
Consider an ε neighbourhood of $(f.g)(x_0)$, write this as $N\left((f.g)(x_0), \varepsilon\right)$. Then we can find ε_1 and ε_2 neighbourhoods of $f(x_0)$ and $g(x_0)$ respectively, such that $N\left(f(x_0), \varepsilon_1\right).N\left(g(x_0), \varepsilon_2\right)$ is contained in $N\left((f.g)(x_0), \varepsilon\right)$ (we haven't proved this yet, just believe it for a few minutes, a proof will follow shortly after the end of this theorem).

Now because f is continuous at x_0, there is some number δ_1 such that for every x in $N(x_0, \delta_1) \cap S$, $f(x)$ will be in $N\left(f(x_0), \varepsilon_1\right)$.
Similarly, because g is continuous at x_0, there is some number δ_2 such that for every x in $N(x_0, \delta_2) \cap T$, $g(x)$ will be in $N\left(g(x_0), \varepsilon_2\right)$.

Now let $N_{x_0} = N(x_0, \delta_1) \cap N(x_0, \delta_2)$.
Then N_{x_0} is a δ-neighbourhood of x_0 such that if x is in $N_{x_0} \cap (S \cap T)$ then $f(x)$ must be in $N\left(f(x_0), \varepsilon_1\right)$ and $g(x)$ must be in $N\left(g(x_0), \varepsilon_2\right)$. This means that $(f.g)(x)$ will be in $N\left((f.g)(x_0), \varepsilon\right)$ as required. Hence $f.g$ is continuous at x_0.

□

Now we need to prove the result we used in the middle of the theorem. The proof that is given is a typical "clever" ε-style proof, where we choose our εs in some strange way, yet at the end of the proof they all work out neatly to give us our result. Of course, when constructing such a proof, we actually begin at the end and work backwards to see what values we need for those εs!

47

Lemma 2.4.1. *If y_1 and y_2 are any real numbers, and we have a $\varepsilon > 0$, then there exists neighbourhoods N_1 of y_1 and N_2 of y_2 such that $N_1.N_2$ is contained in $(y_1 y_2 - \varepsilon, y_1 y_2 + \varepsilon)$*

Proof. Let $N_1 = (y_1 - \varepsilon_1, y_1 + \varepsilon_1)$, where ε_1 is the smaller of

$$\left(\frac{\varepsilon}{3}\right) \text{ and } \left(\frac{\varepsilon}{3|y_2|}\right)$$

Let $N_2 = (y_2 - \varepsilon_2, y_2 + \varepsilon_2)$, where ε_2 is the smaller of

$$1 \text{ and } \left(\frac{\varepsilon}{3|y_1|}\right)$$

Now if z lies in $N_1.N_2$ then:

$$z = (y_1 + k\varepsilon_1)(y_2 + c\varepsilon_2) \quad \text{for some } c \text{ and } k \text{ such that } |c| < 1 \text{ and } |k| < 1.$$

$$= y_1 y_2 + k\varepsilon_1 y_2 + c\varepsilon_2 y_1 + k\varepsilon_1 c\varepsilon_2.$$

Hence we can say that

$$|z - y_1 y_2| = |k\varepsilon_1 y_2 + c\varepsilon_2 y_1 + k\varepsilon_1 c\varepsilon_2|$$
$$\leq \varepsilon_1 |ky_2| + \varepsilon_2 |cy_1| + \varepsilon_1 \varepsilon_2 |ck|, \quad \text{by Triangle Inequality in Section 1.5}$$
$$< \varepsilon_1 |y_2| + \varepsilon_2 |y_1| + \varepsilon_1 \varepsilon_2, \quad \text{since } |c| < 1 \text{ and } |k| < 1$$
$$\leq \frac{\varepsilon}{3} + \frac{\varepsilon}{3} + \frac{\varepsilon}{3}$$
$$= \varepsilon$$

Hence z is in $(y_1 y_2 - \varepsilon, y_1 y_2 + \varepsilon)$ and so $N_1.N_2$ is contained in $(y_1 y_2 - \varepsilon, y_1 y_2 + \varepsilon)$

□

The next theorem is really easy.

Theorem 2.4.3. *Every Polynomial is Continuous*
If $f : \mathbb{R} \longrightarrow \mathbb{R}$ is given by:

$$f(x) = a_0 + a_1 x + a_2 x^2 + a_3 x^3 + \cdots + a_n x^n, \text{ for some positive integer } n$$

then $f :: \mathbb{R} \longrightarrow \mathbb{R}$ is continuous.

Proof. Since the identity function $f(x) = x$ is continuous and the product of two continuous functions is continuous, we know that $f(x) = x^2$ must be continuous.
Hence also $f(x) = x^3, f(x) = x^4, \ldots, f(x) = x^n$ will all be continuous.
We also know that any constant function is continuous and so the products of the powers of x by constants will be continuous.
Lastly the sum of continuous functions is continuous and that gives us our result.

□

Theorem 2.4.4. *The Composition of Two Continuous Functions is Continuous*
If $f : S \longrightarrow \mathbb{R}$ and $g : T \longrightarrow \mathbb{R}$ are two continuous functions such that $g(x)$ is in S for every x in T, then $f \circ g : T \longrightarrow \mathbb{R}$ is continuous.

Proof. Let x_0 be in T, and let $N_{f(g(x_0))}$ be any neighbourhood of $f \circ g(x_0)$.
Then, since f is continuous at $g(x_0)$, there is some neighbourhood $N_{g(x_0)}$ such that if y is in $N_{g(x_0)} \cap S$, then $f(y)$ is in $N_{f(g(x_0))}$.
Then, since g is continuous at x_0, there is some neighbourhood N_{x_0} such that if x is in $N_{x_0} \cap T$, then $g(x)$ is in $N_{g(x_0)}$ so then $f(g(x)) = f \circ g(x)$ is in $N_{f(g(x_0))}$. Hence $f \circ g$ is continuous at x_0.

□

Theorem 2.4.5. *The Reciprocal of a never-zero Continuous Function is Continuous*
If $f : S \longrightarrow \mathbb{R}$ is continuous, then

$$\frac{1}{f} : S - \{f^{-1}(0)\} \longrightarrow \mathbb{R}$$

is continuous.

To avoid any possible confusion, please note that we are using $\frac{1}{f}$ for the reciprocal of f, defined by:

$$\frac{1}{f}(x) = \frac{1}{f(x)}$$

and f^{-1} for the inverse of f defined by:

$$\{f^{-1}(0)\} = \text{the set of points } x \text{ in } S \text{ such that } f(x) = 0$$

They are not the same!!

Before we prove the theorem, let us look at a special case as an example.

Examples 2.4.2
The function $g : \mathbb{R} - \{0\} \longrightarrow \mathbb{R}$ given by $g(x) = \frac{1}{x}$ is continuous.

Let x_0 be a non-zero number, and $N_{g(x_0)}$ any neighbourhood of $g(x_0) = \frac{1}{x_0}$.
Then, since $g(x_0) \neq 0, N_{g(x_0)}$ contains a neighbourhood (a,b) of $g(x_0)$ such that

$$\text{either}$$
$$0 < a < g(x_0) < b \qquad \text{if } g(x_0) > 0$$
$$\text{or}$$
$$a < g(x_0) < b < 0 \qquad \text{if } g(x_0) < 0.$$

Then $N_{x_0} = \left(\frac{1}{b}, \frac{1}{a}\right)$ is a neighbourhood of x_0 and if x is in N_{x_0}, then $g(x)$ is in (a,b) and so is in $N_{g(x_0)}$.

See figure below.
Hence g is continuous.

Now we can prove our theorem.

Proof. Take the composition $g \circ f$ of f and the function $g(x) = \frac{1}{x}$ and use Theorem 2.4.4 to give the result. $\square$

Theorem 2.4.6. *The Quotient of Two Continuous Functions is defined and Continuous on the set where its denominator is non-zero.*
If $f : S \longrightarrow \mathbb{R}$ and $g : T \longrightarrow \mathbb{R}$ are two continuous functions, then $f/g : S \cap T - \{g^{-1}(0)\} \longrightarrow \mathbb{R}$ is continuous.

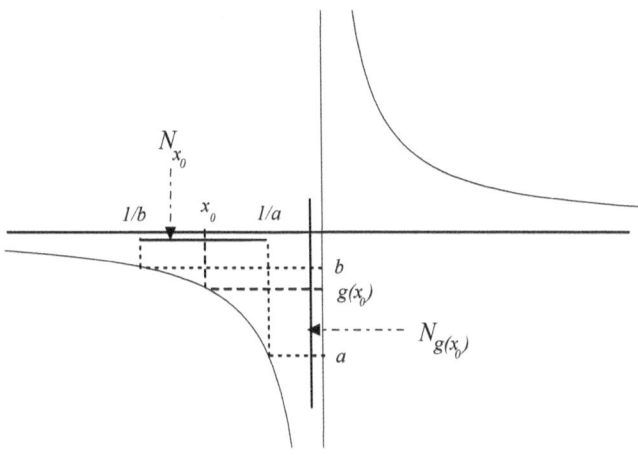

Figure 2.4

Proof. Use the rule for the reciprocal of a function to show that $\frac{1}{g}$ is continuous on $T - \{g^{-1}(0)\}$ and then the rule for the product of two functions to give the result. □

Now we look at what happens when we join together two different continuous functions.

Theorem 2.4.7. Glueing continuous functions together.
Let $f : S \longrightarrow \mathbb{R}$ and $g : T \longrightarrow \mathbb{R}$ be two continuous functions, such that $f(x) = g(x)$ whenever x is in $S \cap T$. Define a new function h by:

$$h(x) = \begin{cases} f(x), & \text{if } x \text{ is in } S \\ g(x), & \text{if } x \text{ is in } T \end{cases}$$

Then $h : S \cup T \longrightarrow \mathbb{R}$ is continuous.

In practice we use this rule most often when $S \cap T$ is just one point, for example:

$$h(x) = \begin{cases} x, & \text{if } x \text{ is in } [0,1] \\ x^2, & \text{if } x \text{ is in } [1,2] \end{cases}$$

Proof. Clearly h is continuous anywhere other than $S \cap T$.
So take any x_0 in $S \cap T$. Then let N be any neighbourhood of $f(x_0) = g(x_0)$. As f and g are continuous at x_0, we can find neighbourhoods N_f and N_g of x_0 such that both $f(N_f)$ and $g(N_g)$ belong to N.
Now let N_{x_0} be defined as $N_f \cap N_g$.
If x belongs to N_{x_0} and also to the domain of h, then $h(x)$ is either $f(x)$ or $g(x)$, and so belongs to N.
Hence $h(N_{x_0})$ belongs to N and so h is continuous at x_0. □

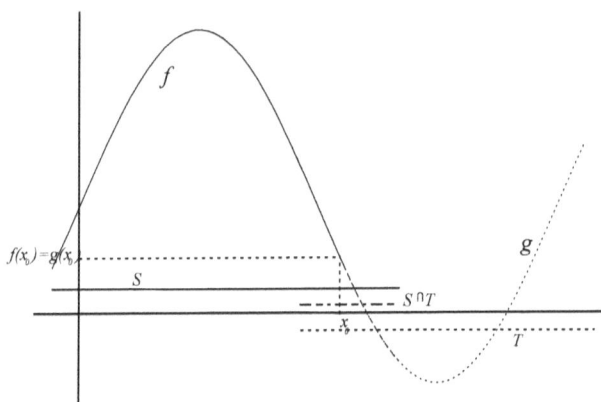

Figure 2.5

Theorem 2.4.8. *The Squeezing Rule.*
Let f, g and $h : S \longrightarrow \mathbb{R}$ be functions, such that $f(x) \leq g(x) \leq h(x)$ whenever x is in S, and $f(a) = h(a)$ for some a in S and f and h are continuous at a, then g is continuous at a.

Proof. Clearly $f(a) = g(a) = h(a)$. For any neighbourhood $N_{g(a)}$ of $g(a)$ we can find a neighbourhood N_a^f of a such that if x is in $N_a^f \cap S$ then $f(x)$ is in $N_{g(a)}$.
Similarly, we can find a neighbourhood N_a^h of a such that if x is in $N_a^h \cap S$ then $h(x)$ is in $N_{g(a)}$.
Define a neighbourhood of a, N_a by $N_a = N_a^f \cap N_a^h$.

If x is in $N_a \cap S$, then $f(x) \leq g(x) \leq h(x)$ and $f(x)$ and $h(x)$ are both in $N_{g(a)}$. So $g(x)$ must be in $N_{g(a)}$ and g is continuous at a.
□

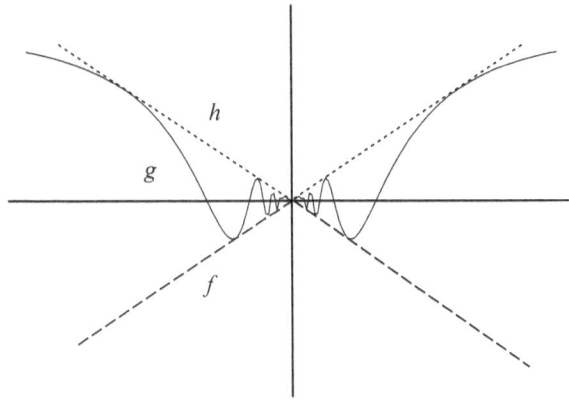

Figure 2.6

2.5. Completeness Properties of Continuous Functions

Although the theorems given in the previous section were all about functions from subsets of $\mathbb{R}$ to $\mathbb{R}$, in fact they would have worked, for instance, just as well for functions from subsets of $\mathbb{Q}$ to $\mathbb{Q}$, or indeed any other ordered field F.

In this section, however, we will be making use of that *essential* property of the real numbers, the completeness property. Since that was all about sets being bounded above and/or below and the bounds of such sets, it should be no surprise that we begin by extending those ideas to functions.

The definition that follows just gives the obvious meanings to the words we have already met for numbers, supremum, infimum, bounds, etc. to functions by applying them to the sets of values that a function generates.

Definition 2.5.1. *The function $f : S \longrightarrow \mathbb{R}$ is <u>bounded above</u> if the set of its values, $f(S)$, is bounded above.*
The function $f : S \longrightarrow \mathbb{R}$ is <u>bounded below</u> if the set of its values, $f(S)$, is bounded below.
The function $f : S \longrightarrow \mathbb{R}$ is <u>bounded</u> if it is bounded above and bounded below.
For such a function, f, we denote its <u>supremum</u> by $\sup f$ or $\sup_{x \text{ in } S} f(x)$ or $\sup\{f(x) : x \text{ in } S\}$
We denote its <u>infimum</u> by $\inf f$ or $\inf_{x \text{ in } S} f(x)$ or $\inf\{f(x) : x \text{ in } S\}$
The supremum and infimum of a function are called its <u>bounds</u>.

If there is some value of x in S such that $f(x) = \sup f$ then we say that f attains its supremum. And, of course, if there is some value of x in S such that $f(x) = \inf f$ then we say that f attains its infimum.

Examples 2.5.1
1. The function $f(x) = x^2$, defined on the whole of $\mathbb{R}$, is bounded below but not bounded above. It attains its infimum.

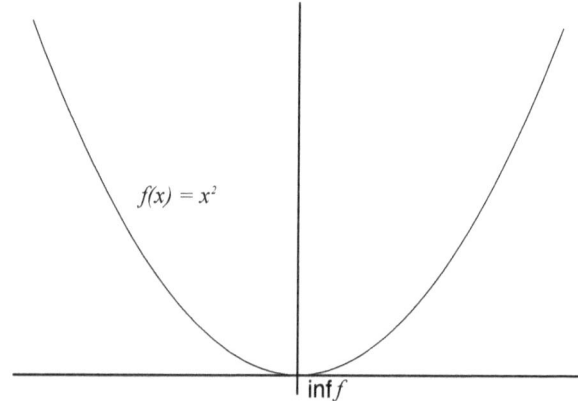

Figure 2.7

2. The function $f(x) = 1/x$, defined on $(0, 1]$ is bounded below but not above. It attains its infimum.

2.5. COMPLETENESS PROPERTIES OF CONTINUOUS FUNCTIONS

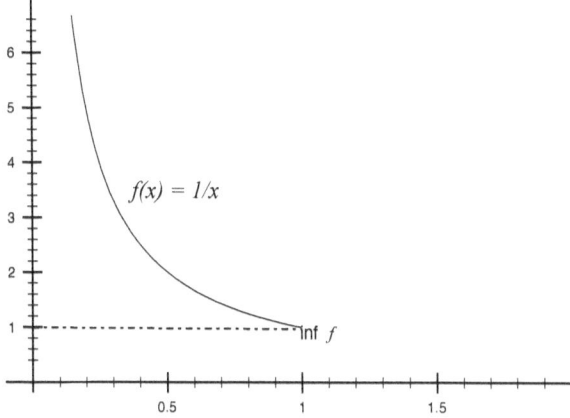

Figure 2.8

3. The function $f(x) = x^2$, defined on $(0, 2)$ is bounded above and below, but does not attain its bounds.

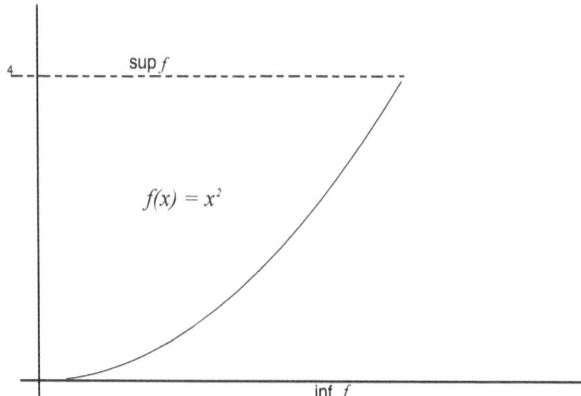

Figure 2.9

CHAPTER 2. CONTINUITY

There now follows four especially important theorems describing continuity for a function defined on an interval of the reals.

Theorem 2.5.1. *If $f : I \longrightarrow \mathbb{R}$ is continuous and I is a closed bounded interval of $\mathbb{R}$, then f is bounded and attains its bounds.*

Proof. First of all, I will prove that if f is continuous at x_0 in I then there is a neighbourhood of x_0 on which f is bounded.
To show this, take a neighbourhood of $f(x_0)$ of radius 1, $N(f(x_0), 1)$. Then, because f is continuous at x_0, there is a neighbourhood of x_0, N_{x_0}, such that $f(N_{x_0} \cap I)$ belongs to $N(f(x_0), 1)$.
But this means that for all x in N_{x_0}, the value of $f(x)$ lies between $f(x_0) + 1$ and $f(x_0) - 1$, hence f is bounded on that neighbourhood.

Now we are going to use the Creeping Lemma, Theorem 1.5.3 to prove that f is bounded.
Define a relation ρ on $I = [a, b]$ by:

$$u\rho v \text{ if and only if } f \text{ is bounded on } [u, v]$$

(a) ρ is transitive - left as an exercise for you.
(b) Now let x_0 belong to I. We have already shown that there is a neighbourhood N_{x_0} of x_0 on which f is bounded, hence for any two numbers u, v in N_{x_0}, f will be bounded on $[u, v]$, hence $u\rho v$ for all such u and v.
The Creeping Lemma now tells us that $a\rho b$, in other words f is bounded on $[a, b] = I$.

Now we shall show that f actually attains its bounds.
I am only going to show that f attains its upper bound, the proof that it attains its lower bound is left to you as an exercise. (It is virtually identical to the proof for the upper bound!)
Let $m = \sup f$. Then for all x in I, $f(x) \leq m$.
Let us suppose that f does **not** attain its supremum. (*)
That means that $f(x) < m$ for all x in I. In other words $m - f(x) \neq 0$ for all values of x in I.

Now consider the function $m - f : I \longrightarrow \mathbb{R}$. This is the difference of two continuous functions hence is also a continuous function. It is also never zero. Hence from Theorem 2.4.5 we know that

$$\frac{1}{m-f} : I \longrightarrow \mathbb{R}$$

is continuous. But $\frac{1}{m-f}$ is defined on a closed interval so is bounded, by what we have already proved.
Hence we have some number k, with $k > 0$, such that:

$$\frac{1}{m-f(x)} \leq k \qquad \text{for all } x \text{ in } I$$

If we rearrange this we get:

$$m - f(x) \geq 1/k \qquad \text{for all } x \text{ in } I$$

Thus:

$$f(x) \leq m - 1/k \qquad \text{for all } x \text{ in } I$$

This last statement contradicts the definition of m as the supremum of f. So our supposition (*) must be wrong and so f must attain its upper bound.

□

Now we are going to prove a result that seems obvious and is very important, yet it does need to be proved properly.

CHAPTER 2. CONTINUITY

Theorem 2.5.2. *The Intermediate Value Theorem*
If $f : I \longrightarrow \mathbb{R}$ is continuous, and I is an interval, then given any two numbers a and b in I and any number z between $f(a)$ and $f(b)$, then there is some c between a and b such that $f(c) = z$.

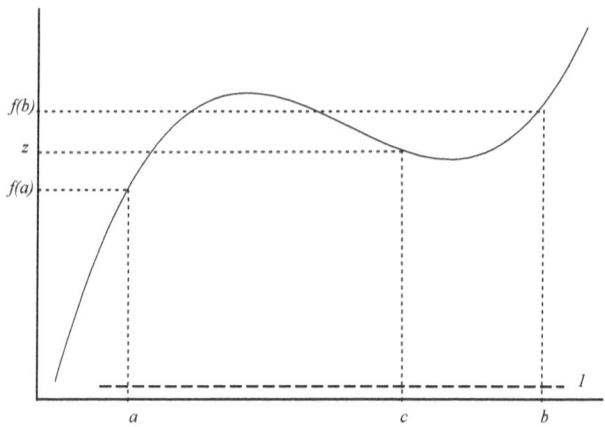

Figure 2.10

Proof. We will assume that a is the left-hand point and b the right-hand point, so $a < b$.
Furthermore we assume that $f(a) < f(b)$, so we can say that $f(a) < z < f(b)$
Now we replace f by g defined by $g(x) = f(x) - z$.
 Putting all these assumptions together, we have that $g(a) < 0$ and $g(b) > 0$ and we are trying to prove that that there is some point c in (a,b) where $g(c) = 0$.

 Let us assume that this is false, in other words, $g(x) \neq 0$ for all values of x in $[a,b]$.
We will use the Creeping Lemma to prove this is impossible.
 Define a relation ρ on $[a,b]$ by $u\rho v$ if and only if $g(u)$ and $g(v)$ are of the same sign.
Clearly ρ is transitive.
Let x_0 be a point in $[a,b]$. So $g(x_0) \neq 0$ and thus there is a neighbourhood, $N_{g(x_0)}$, of $g(x_0)$ which does not contain 0. This means that

there is a neighbourhood, N_{x_0}, of x_0 such that if x is in $N_{x_0} \cap I$, then $g(x)$ is in $N_{g(x_0)}$ and so $g(x)$ has the same sign as $g(x_0)$. So $u\rho v$ for all u and v in N_{x_0}.

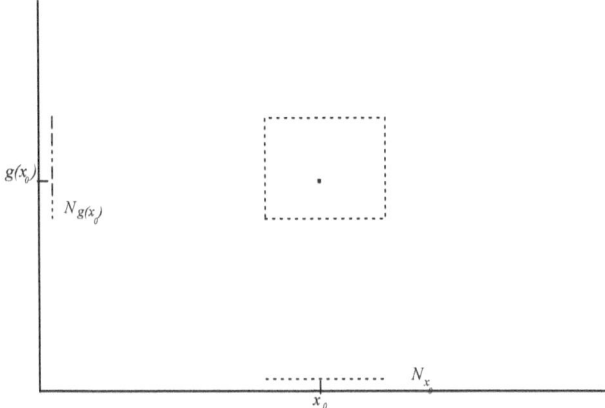

Figure 2.11

So ρ satisfies the conditions of the Creeping Lemma, Theorem 1.5.3, hence $a\rho b$, that is $g(a)$ and $g(b)$ have the same sign. But this contradicts our statement above that $g(a) < 0$ and $g(b) > 0$. □

Theorem 2.5.3. *If $f : I \longrightarrow \mathbb{R}$ is continuous, and I is an interval, then $f(I)$ is an interval.*

This is an immediate consequence of the Intermediate Value Theorem.

Theorem 2.5.4. *If $f : I \longrightarrow \mathbb{R}$ is continuous, and I is a closed bounded interval of $\mathbb{R}$, then $f(I)$ is a closed bounded interval.*

Proof. $f(I)$ is an interval by the previous theorem, and is bounded and attains its bounds by Theorem 2.5.1.
Since it attains its bounds, the interval is closed. □

Now we have proved all the results about continuous functions that we need and will now turn our attention to differentiation, the second major topic in Analysis.

2.6. Exercises

1. Suppose that $f : S \longrightarrow \mathbb{R}$ and $g : T \longrightarrow \mathbb{R}$ are two continuous functions and that we have an ε neighbourhood of $(f+g)(x_0)$, $N\left((f+g)(x_0), \varepsilon\right)$, so $N\left(f(x_0), \frac{1}{2}\varepsilon\right)$ and $N\left(g(x_0), \frac{1}{2}\varepsilon\right)$ are neighbourhoods of $f(x_0)$ and $g(x_0)$ respectively.
Using the definition of $A+B$ in Chapter 1, prove that $N\left(f(x_0), \frac{1}{2}\varepsilon\right) + N\left(g(x_0), \frac{1}{2}\varepsilon\right)$ will be contained in $N\left((f+g)(x_0), \varepsilon\right)$.

2. If $f : S \longrightarrow \mathbb{R}$ and $g : T \longrightarrow \mathbb{R}$ are two continuous functions such that if both f and g are continuous at x_0, then $f+g$ is continuous there also, prove that $f+g : S \cap T \longrightarrow \mathbb{R}$ is continuous.

3. If $f : S \longrightarrow \mathbb{R}$ and $g : T \longrightarrow \mathbb{R}$ are two continuous functions, such that if both f and g are continuous at x_0, then $f.g$ is continuous there also, prove that $f.g : S \cap T \longrightarrow \mathbb{R}$ is continuous.

4. If $f : I \longrightarrow \mathbb{R}$ is continuous and I is a closed bounded interval of $\mathbb{R}$, then f is bounded and attains its lower bound.

5. Consider the following functions $f : (-2, 3) \longrightarrow \mathbb{R}$
    (a)
    $$f(x) = \frac{x+1}{2x-7}$$
    (b)
    $$f(x) = |x-1|$$
    (c)
    $$f(x) = \begin{cases} 1, & \text{if } x < 1 \\ 3x, & x \geq 1 \end{cases}$$
    (d)
    $$f(x) = \begin{cases} 1, & \text{if } x \text{ is in } [0,1] \\ 0, & \text{for all other values of } x \end{cases}$$
    (e)
    $$f(x) = \frac{x^2+9}{x^2-9}$$

(f)
$$f(x) = \begin{cases} x(x-2), & \text{if } x \text{ is in } (0,2) \\ 0, & \text{for all other values of } x \end{cases}$$

For each function determine if it is:-
(i) continuous on (-2,3)
(ii) continuous on [0,1]
(iii) bounded on (-2,3)
    and in the cases where the function is bounded, whether it:-
(iv) attains its supremum
(v) attains its infimum

Chapter 3
Differentiation - Basic Definitions

3.1. Objectives

The aim of this chapter is to introduce the ideas of differentiation in a more formal way than you will have met before. By the end of the chapter you should be able to differentiate a function from the basic definition and appreciate the difference between the differential coefficient and the derivative of a function.

3.2. New terms and symbols introduced

chord slope function	$p_{x_0}(x)$
differentiable at x_0	
differentiable	
differential coefficient of f at x_0	$f'(x_0), Df(x_0)$
derivative of f	f', Df
second derivative of f	$f'', D^2 f$

3.3. What is differentiation?

Differentiation is one of the two major concepts of analysis. With integration it gives analysis its particular flavour. Of course it is true that without functions, continuity, limits and the other concepts we will look at, analysis would be impossible, yet it is from differentiation and integration that all the really exciting ideas arise.

A certain amount of the interest is due to the close connection between the mathematics and certain ideas in physics and you are encouraged to look at the application of mathematical ideas to physical problems, which is outside the scope of this book. However, initially we shall define our ideas in precise mathematical form and discuss their significance in terms of some important mathematical problems.

We have already restricted our attention to continuous functions as these ''well-behaved'' and obtained a number of results about them;

but when we further restrict attention to functions that are "very well-behaved" we will obtain the most interesting and powerful results.

In order to see the kind of misbehaviour that continuous functions are capable of, look at the following diagrams (adapted from [Spi06]):-

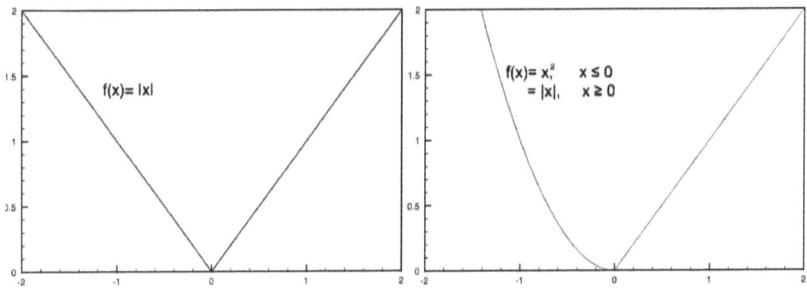

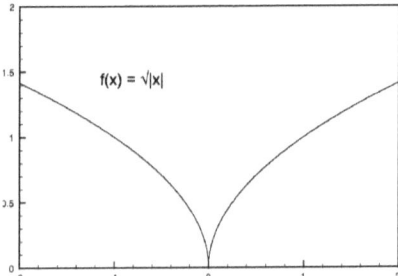

Figure 3.1: Continuous but "bent" at (0,0)

You can see that each graph has a sharp point at the origin, (0,0). Everywhere else the graph is smooth. How can we characterize this? Well, everywhere except at those sharp pointed origins the curve (or straight line) has a tangent, a line lying alongside the original and pointing in the same direction as that original. But at the origin there is a sudden and large change of direction, so in each of the graphs in Figure 3.1 it is not possible to draw a "tangent line" at the point (0,0). The graphs are too "bent" at that point. We are using the inverted commas as we have not defined what "tangent line" or "bent" means.

3.3. WHAT IS DIFFERENTIATION?

Let us look more closely at the idea of "tangent line".

It isn't possible to define a "tangent line" by saying it is a line that only meets the graph once, one of the more common explanations given of a tangent, as then the line shown in Figure 3.2 wouldn't be a tangent and the parabola in Figure 3.3 would have two tangents at every point!

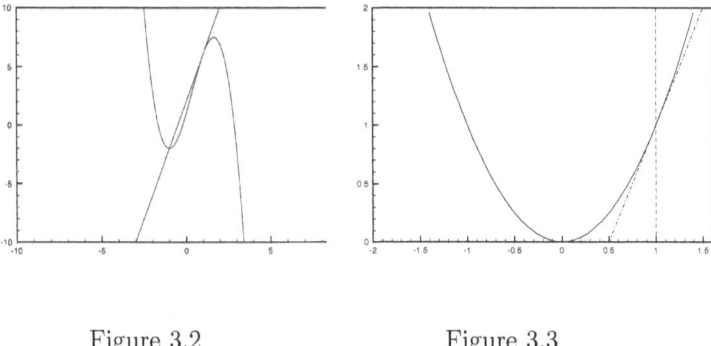

Figure 3.2 Figure 3.3

A more promising approach would be to regard the "tangent line" as a linear approximation to the original function. By this we mean that we can construct a linear function $y = mx + c$ for suitable values of m and c that approximates the original function when we are near to the value of x in which we are interested.

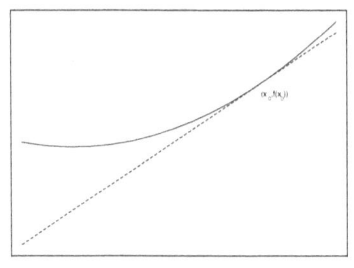

Figure 3.4

A moment's thought should convince you that the equation of our "linear approximation" will be $y = f(x_0) + (x - x_0)\lambda$ where λ is the slope of the line. So this will be the equation of our "tangent line". [If you are not happy about this statement, stop here and prove it for yourself. Hint: translate the origin to $(x_0, f(x_0))$].

By the tangent line being a linear approximation to the function f,

we mean that:-
$$f(x) \simeq f(x_0) + (x - x_0)\lambda \text{ for } x \text{ close to } x_0$$

This suggests that we concentrate our attention on finding a "λ" to make that equation true.

The obvious place to start is by constructing a straight line through the point $(x_0, f(x_0))$ which meets the graph of the function in another point close by. This is called a "chord" for the function.

Formally:-
If $f : A \longrightarrow \mathbb{R}$ is a continuous function then define another function $p_{x_0} : A \longrightarrow \mathbb{R}$ by
$$f(x) = f(x_0) + (x - x_0)p_{x_0}(x)$$

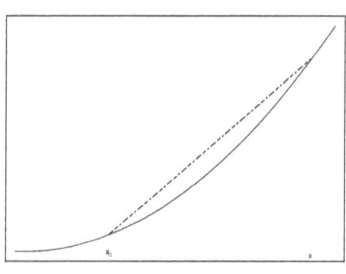

Figure 3.5

If $x \neq x_0$ then this determines the value of p_{x_0} to be given by the equation:
$$p_{x_0}(x) = \frac{f(x) - f(x_0)}{x - x_0}$$

and thus p_{x_0} is precisely the slope of the chord from $(x_0, f(x_0))$ to $(x, f(x))$. Hence we call p_{x_0} the chord slope function at x_0

At $x = x_0$, the value of $p_{x_0}(x)$ is not determined by the above equation, so we must specify it separately. In order to see how we might do this, let us look at some examples.

Examples 3.3.1
1. Suppose $f(x) = c$, a constant, then $p_{x_0}(x) = 0$ if $x \neq x_0$, as $f(x) = f(x_0) = c$; so the chord joining (x_0, c) to (x, c) always has slope 0.

It seems sensible to choose $p_{x_0}(x_0) = 0$.

2. Suppose $f(x) = x$ then $p_{x_0} = \frac{f(x) - f(x_0)}{x - x_0} = \frac{x - x_0}{x - x_0} = 1$, if $x \neq x_0$; i.e the chord from (x_0, x_0) to (x, x) always has slope 1.

3.3. WHAT IS DIFFERENTIATION?

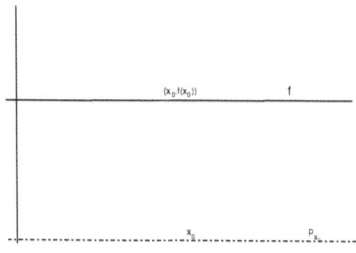

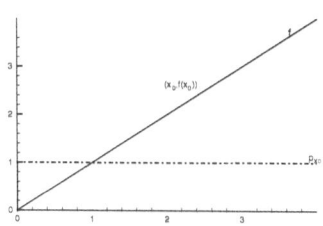

It seems sensible to choose $p_{x_0}(x_0) = 1$.

3. Suppose $f(x) = x^2$ then $p_{x_0} = \frac{f(x)-f(x_0)}{x-x_0} = \frac{x^2-x_0^2}{x-x_0} = x + x_0$, if $x \neq x_0$.

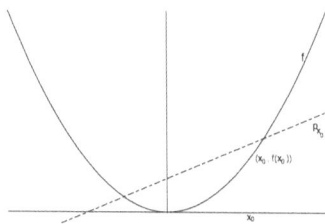

It seems sensible to choose $p_{x_0}(x_0) = 2x_0$.

4. Suppose $f(x) = |x|$, then there are three different cases to consider:

(a) <u>$x_0 > 0$</u> then $p_{x_0}(x) = \frac{|x|-|x_0|}{x-x_0} = \begin{cases} 1, & \text{if } x > 0, x \neq x_0 \\ \frac{-(x+x_0)}{x-x_0}, & \text{if } x < 0 \end{cases}$

67

CHAPTER 3. DIFFERENTIATION - BASIC DEFINITIONS

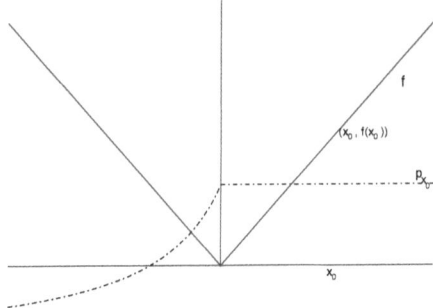

So it seems sensible to choose $p_{x_0}(x_0) = 1$.

(b) $x_0 < 0$ This case is similar to the previous case and hence we would choose $p_{x_0}(x_0) = -1$ [You might like to try working through this case yourself].

(c) $x_0 = 0$ In this case,
$$p_0(x) = \frac{|x|}{x} = \begin{cases} 1, & \text{if } x > 0 \\ -1, & \text{if } x < 0 \end{cases}$$

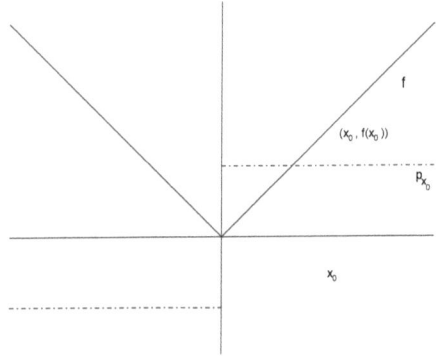

Here there is no sensible choice for $p_0(0)$.

5. Suppose f is a step function, e.g $f(x) = \begin{cases} 1, & \text{if } x < x_0 \\ 2, & \text{if } x \geq x_0 \end{cases}$

then $p_{x_0}(x) = \frac{f(x)-2}{x-x_0} = \begin{cases} 0, & \text{if } x > x_0 \\ \frac{-1}{x-x_0}, & \text{if } x < x_0 \end{cases}$

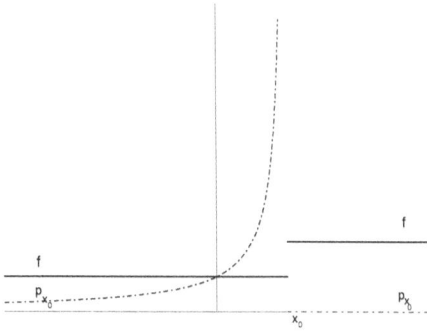

Again there is no sensible choice for $p_{x_0}(x_0)$.

We can now draw a few conclusions from the above five examples.

Firstly the choice of a sensible value for $p_{x_0}(x_0)$ was, in fact just choosing a value to make p_{x_0} continuous at x_0.
Secondly, this value $p_{x_0}(x_0)$ is just the slope of the tangent to the graph of the function at $(x_0, f(x_0))$.
Thirdly if we have corners or steps in our graph, there is no value of $p_{x_0}(x_0)$ which will make p_{x_0} continuous at x_0.

So it looks as if we have found the mechanism for distinguishing between "well-behaved" and "very well-behaved" functions.

We are now in a position to make the following definitions:

Definition 3.3.1. *Let $f : A \longrightarrow \mathbb{R}$ be a continuous function, and $x_0 \in a$.*
1. f is <u>differentiable at x_0</u> if there exists a function $p_{x_0} : A \longrightarrow \mathbb{R}$ such that
$f(x) = f(x_0) + (x - x_0)p_{x_0}(x)$, for all $x \in A$ and p_{x_0} is continuous at x_0.
2. If f is differentiable at every x_0 in A then we say f is <u>differentiable on A</u>.
3. The number $p_{x_0}(x_0)$ is called the <u>differential coefficient of f at x_0</u>,

and written $f'(x_0)$ or $Df(x_0)$.
4. If f is differentiable on A then the <u>derivative of f</u> is the function, written f' or Df, with domain A, defined by

$$Df(x) = f'(x) = p_x(x) \text{ for all } x \in A$$

<u>Note</u> The distinction between the derivative of a function and its differential coefficient is exactly analagous to the distinction between a function and its value at a particular point. It is for this reason that it is essential to use the symbol Df of f', for the derivative rather than the Leibnizian alternative $\frac{df}{dx}$.

We can now illustrate our definitions by some examples.

Examples 3.3.2
1. Suppose $f(x) = c$, then from our previous examples, number 1, $f'(x_0) = 0$.
So the derivative is the zero function, i.e., $f : x \mapsto 0$, for all values of x in $\mathbb{R}$.

2. Suppose $f(x) = x$, then from number 2 of the previous examples, $Df(x_0) = 1$.
So the derivative is the function: $Df : x \mapsto 1$, for all values of x in $\mathbb{R}$.

3. Suppose $f(x) = x^2$, then using number 3 of the previous examples, $Df(x_0) = 2x_0$.
So the derivative is the function: $Df : x \mapsto 2x$, for all values of x in $\mathbb{R}$.

4. Let $f(x) = x^3$. We will have to work this one out for ourselves.

$$\begin{aligned}f(x) &= x^3 \\ &= x_0^3 + (x^3 - x_0^3), \text{this is a standard pice of low cunning-} \\ &\qquad \text{adding and subtracting the same number!} \\ &= x_0^3 + (x - x_0)(x^2 + xx_0 + x_0^2) \\ &= f(x_0) + (x - x_0)p_{x_0}(x)\end{aligned}$$

where $p_{x_0}(x_0) = x^2 + xx_0 + x_0^2$
Now p_{x_0} is a polynomial, hence it is continuous for all values of x
So $p_{x_0}(x_0) = 3x_0^2$

Therefore $f'(x_0) = 3x_0^2$
So $f' : x \mapsto 3x^2$ is the derivative of $f(x) = x^3$.

5. Suppose $f(x) = \frac{1}{x^2}$
Then
$$f(x) = \frac{1}{x^2}$$
$$= \frac{1}{x_0^2} + (\frac{1}{x^2} - \frac{1}{x_0^2}), \text{ a good trick is worth using again}$$
$$= \frac{1}{x_0^2} + (\frac{x_0^2 - x^2}{x^2 x_0^2})$$
$$= \frac{1}{x_0^2} + (x - x_0)\frac{-x_0 + x}{x^2 x_0^2}$$
$$= f(x_0) + (x - x_0)p_{x_0}(x)$$

where $p_{x_0}(x_0) = -\frac{(x_0+x)}{x^2 x_0^2}$ and is continuous as long as $x \neq 0$.
Hence f is differentiable and $Df(x_0) = -\frac{2x_0}{x_0^4} = \frac{-2}{x_0^3}$

$$\text{So } Df : x \mapsto \frac{-2}{x^3}$$

6. Let $f(x) = \frac{1}{x}$
Try this one yourself, it is not too different from example 5. If you get stuck you will find the complete solution written out in Section 5.2, page 81.

7. Let $f(x) = \sqrt{x}$

$$f(x) = x^{1/2}$$
$$= x_0^{1/2} + x^{1/2} - x_0^{1/2}$$
$$= x_0^{1/2} + (x^{1/2} - x_0^{1/2})\frac{x^{1/2} + x_0^{1/2}}{x^{1/2} + x_0^{1/2}} \quad \text{(another standard piece of low}$$

cunning -
dividing and multiplying
by the same non-zero number
does not change the value).

$$= x_0^{1/2} + (x - x_0)\frac{1}{x^{1/2} + x_0^{1/2}} \quad \text{(the difference of two squares)}$$
$$= f(x_0) + (x - x_0)p_{x_0}(x)$$

with $p_{x_0}(x) = \frac{1}{x^{1/2}+x_0^{1/2}}$, which is continuous for all $x \geq 0$.

So f is differentiable and $f'(x_0) = \frac{1}{2x_0^{1/2}}$
thus $f' : x \mapsto \frac{1}{2x^{1/2}}$ is the derivative of f.

There is little point in looking at further problems involving negative or fractional indices as they become more a test of algebra than analysis! Anyway, we shall derive a general formula for such expressions in Section 3. We can start now by proving the formula for the derivative of the function $f(x) = x^n$ where n is a positive integer.

Theorem 3.3.1. *If $f(x) = x^n$ then $f'(x) = nx^{n-1}$.*

Proof.
$$f(x) = x^n = x_0^n + x^n - x_0^n = x_0^n + (x - x_0)(x^{n-1} + x_0 x^{n-2} + \cdots + x_0^{n-1})$$

(Here we are just using the standard algebraic factorization).
So,
$$f(x) = f(x_0) + (x - x_0)p_{x_0}(x)$$
with
$$p_{x_0}(x) = x^{n-1} + x_0 x^{n-2} + \cdots + x_0^{n-1}$$

Being a polynomial, $p_{x_0}(x)$ is continuous for all values of x.
Hence
$$f'(x_0) = p_{x_0}(x_0) = nx_0^{n-1}$$

Therefore
$$f' : x \mapsto nx^{n-1}$$

$\square$

For any function f, we obtain, by taking the derivative, a new function Df. The idea of differentiability can then be applied to the function Df, giving another function $D(Df)$, usually written $D^2 f$ or f'' and called the <u>second derivative of f</u>.

There is no reason for stopping at the second derivative, we can go on differentiating as far as we like getting $D^3 f, D^4 f$ and so on. These functions are called <u>higher order derivatives of f</u>.

3.4. Exercises

1. Prove, starting from the definition (and drawing a picture to illustrate):
(a) if $g(x) = f(x) + c$ then $g'(x) = f'(x)$.
(b) if $g(x) = c.f(x)$ then $g'(x) = c.f'(x)$.

2. Let $f(x) = x^3$, if x is rational and $f(x) = 0$ if x is irrational. Prove that f is differentiable at 0. [Don't be frightened by this peculiar looking function. Just write out the definition of $Df(0)$ and see what you get.]

3. (a) Suppose that we have three functions, $f, g,$ and h defined as follows:-
$f(a) = g(a) = h(a)$ and $f(x) \leq g(x) \leq h(x)$ for all x. Also suppose that $f'(a) = h'(a)$.
Prove that g is differentiable at a and that $f'(a) = g'(a) = h'(a)$.
Hint: start by finding $g'(a)$
(b) Find an example to show that if we omit the hypothesis that $f(a) = g(a) = h(a)$ the conclusion is no longer true.

4. (a) Suppose that $g(x) = f(x+c)$. Prove that $Dg(x) = Df(x+c)$.
(b) If $g(x) = f(cx)$ then show that $Dg(x) = c.Df(x)$.
(c) Suppose that f is a differentiable function that is also periodic, that is there is some fixed number a such that $f(x+a) = f(x)$, for all values of x. (a is called the period of f). Prove that Df is also periodic.

Chapter 4
Elementary properties of differentiable functions:1

4.1. Objectives

This chapter gives and proves the rules for differentiating the sum, product and composition of two functions. By the end of the chapter not only should you know the rules and how to use them but you should have an appreciation of the proofs.

4.2. The sum and product of two functions

To enable us to differentiate functions without always going back to the basic definition we use a number of rules, the first two are in this section the others will be in the next.

Theorem 4.2.1. *If $f : A \longrightarrow \mathbb{R}$ and $g : A \longrightarrow \mathbb{R}$ are differentiable then $f + g : A \longrightarrow \mathbb{R}$ is differentiable and $D(f + g) = Df + Dg$*

Briefly:-
The sum of two differentiable functions is differentiable and $D(f+g) = Df + Dg$

In order to prove this theorem we need to use the result that the sum of two continuous functions is continuous.

Proof. Let $x_0 \in A$. Since both f and g are differentiable at x_0, there exist functions p_{x_0} and q_{x_0} both continuous at x_0 such that:

$$f(x_0) = f(x_0) + (x - x_0)p_{x_0}(x) \text{ and}$$
$$g(x_0) = g(x_0) + (x - x_0)q_{x_0}(x)$$

Adding the two lines together we get:

$$\begin{aligned}(f+g)(x) &= f(x) + g(x) \\ &= f(x_0) + g(x_0) + (x-x_0)[p_{x_0}(x) + q_{x_0}(x)] \\ &= (f+g)(x_0) + (x-x_0)r_{x_0}(x)\end{aligned}$$

where $r_{x_0}(x) = p_{x_0}(x) + q_{x_0}(x)$

As the sum of two continuous functions is continuous, it follows that r_{x_0} is continuous at x_0, hence $f + g$ is differentiable there, and as x_0 was any point in A, we can conclude that $f + g$ is differentiable. Further we have also shown that

$$D(f+g)(x_0) = r_{x_0}(x_0) = p_{x_0}(x_0) + q_{x_0}(x_0) = Df(x_0) + Dg(x_0)$$

So

$$D(f+g) = Df + Dg$$

$\square$

The rule for a product is not quite so simple yet it is pleasantly symmetric.

Theorem 4.2.2. *If $f : A \longrightarrow \mathbb{R}$ and $g : A \longrightarrow \mathbb{R}$ are differentiable then $f.g : A \longrightarrow \mathbb{R}$ is differentiable and $D(f.g) = f.Dg + g.Df$*

Briefly:-
The product of two differentiable functions is differentiable and $D(f.g) = f.Dg + g.Df$

In order to prove this theorem we need to use the result that the product of two continuous functions is continuous.

Proof. This proof starts the same as that of the previous theorem. In fact the first four lines are the same:-
Let $x_0 \in A$. Since both f and g are differentiable at x_0, there exist functions p_{x_0} and q_{x_0} both continuous at x_0 such that:

$$\begin{aligned}f(x_0) &= f(x_0) + (x-x_0)p_{x_0}(x) \text{ and} \\ g(x_0) &= g(x_0) + (x-x_0)q_{x_0}(x)\end{aligned}$$

But now we will multiply the previous two lines together to give us:-

$$(f.g)(x) = f(x).g(x)$$
$$= [f(x_0) + (x - x_0)p_{x_0}(x)][g(x_0) + q_{x_0}(x)]$$
$$= (f.g)(x_0) + (x - x_0)[p_{x_0}(x)g(x_0) + f(x_0)q_{x_0}(x)$$
$$+ (x - x_0)p_{x_0}(x)q_{x_0}(x)]$$

Let $r_{x_0}(x) = p_{x_0}(x)g(x_0) + f(x_0)q_{x_0}(x) + (x - x_0)p_{x_0}(x)q_{x_0}(x)$
As it is a sum of products of continuous functions, r_{x_0} will be continuous at x_0, thus $f.g$ is differentiable there, hence on the whole of A and

$$D(f.g)(x_0) = r_{x_0}(x_0) = Df(x_0).g(x_0) + f(x_0)).Dg(x_0)$$

So:

$$D(f.g) = f.Dg + g.Df$$

$\square$

We are now able to differentiate several new functions, for instance if $g(x) = c.f(x)$ then $g'(x) = c.f'(x)$. [Try to prove this for yourself, use the product rule with one of the functions being the constant function.]

If $f(x) = a_n x^n + a_{n-1} x^{n-1} + \cdots + a_2 x^2 + a_1 x + a_0$
then $f'(x) = na_n x^{n-1} + (n-1)a_{n-1} x^{n-2} + \cdots + 2a_2 x + a_1$

If we have more than two functions multiplied together we can extend Theorem 2.2. It is an easy exercise to derive a formula for the product of three functions and even to prove (by induction) the formula for the product of n functions:-

$$D(f_1.f_2.....f_n)(x) = \sum_{i=1}^{n} f_1(x) \ldots f_{i-1}(x) Df_i(x) f_{i+1}(x) \ldots f_n(x)$$

4.3. The composition of two functions

In order to be able to differentiate any more interesting functions, we will need a rule for the derivative of the composition of two functions, $f \circ g$. It seems reasonable to presume that for $f \circ g$ to be differentiable at x_0, g would have to be differentiable at x_0 and f would have to be differentiable at $g(x_0)$.

In fact we will prove this to be the case in the following theorem, known as the Chain Rule:-

Theorem 4.3.1. *If $f : A \longrightarrow \mathbb{R}$ and $g : B \longrightarrow \mathbb{R}$ are differentiable and $\forall x \in B, g(x) \in A$, then $f \circ g : B \longrightarrow \mathbb{R}$ is differentiable and $D(f \circ g) = (Df \circ g).Dg$*

The nub of this proof will be that the composition of two continuous functions is continuous.

Proof. Let x_0 be any point in the set B. Since g is differentiable, there exists a function $q_{x_0} : B \longrightarrow \mathbb{R}$ such that

$$g(x) = g(x_0) + (x - x_0)q_{x_0}(x), \forall x \in B \qquad (4.1)$$

and q_{x_0} is continuous at x_0.

Since f is differentiable at $g(x_0)$, there exists a function $p_{g(x_0)} : A \longrightarrow \mathbb{R}$, continuous at $g(x_0)$, such that

$$f(y) = f(g(x_0)) + (y - g(x_0))p_{g(x_0)}(y), \forall y \in A \qquad (4.2)$$

Now if we restrict our attention to one particular subset of A, namely $g(B)$, we can write $y = g(x)$ for some $x \in B$ for every y in that subset $g(B)$.
So equation (4.2) becomes

$$f(g(x)) = f(g(x_0)) + (g(x) - g(x_0))p_{g(x_0)}(g(x))$$

We can now substitute for $g(x) - g(x_0)$ from equation (4.1) and also write $(f \circ g)(x)$ for $f(g(x))$ to get:

$$(f \circ g)(x) = (f \circ g)(x_0) + (x - x_0)q_{x_0}(x)p_{g(x_0)}(g(x)) \qquad (4.3)$$

which we can also write as

$$(f \circ g)(x) = (f \circ g)(x_0) + (x - x_0)(p_{g(x_0)} \circ g)(x).q_{x_0} \qquad (4.4)$$

Since $p_{g(x_0)} \circ g$ is the composition of two continuous functions it will be continuous at x_0 and its value there will be

$$(p_{g(x_0)} \circ g)(x_0) = p_{g(x_0)}(g(x_0)) = Df(g(x_0)) = (Df \circ g)(x_0) \qquad (4.5)$$

q_{x_0} is continuous at x_0 and its value there is $Dg(x_0)$, so the product $(p_{g(x_0)}).q_{x_0}$ is also continuous at x_0, with value there of

$$D(f \circ g)(x_0).Dg(x_0) \tag{4.6}$$

Thus we have found a function $r_{x_0} : B \longrightarrow \mathbb{R}$, continuous at x_0 such that
$$(f \circ g)(x) = (f \circ g)(x_0) + (x - x_0)r_{x_0}(x) \tag{4.7}$$
So $f \circ g$ is differentiable at x_0 and hence on the whole of B and

$$D(f \circ g)(x_0) = r_{x_0}(x_0) = [(Df \circ g).Dg](x_0)$$

□

4.4. Exercises

1. Extend the product rule, Theorem 4.2.2, to the case of the product of three functions, f, g and h.

2. Use induction to prove the product formula for n functions, namely:-

$$D(f_1.f_2.....f_n)(x) = \sum_{i=1}^{n} f_1(x)\ldots f_{i-1}(x)Df_i(x)f_{i+1}(x)\ldots f_n(x)$$

3. Use the Chain Rule to find $Df(x)$ for each of the following:-
(a) $f(x) = (x + x^2)^3$
(b) $f(x) = (((x^2 + x)^3) + x)^4) + x)^5$

Chapter 4. Elementary Properties of Differentiable Functions:1

Chapter 5
Elementary properties of differentiable functions:2

5.1. Objectives

In this chapter we meet the remaining rules for differentiating a function, viz. the cases of the reciprocal of a function, the quotient of two functions and the inverse of a function.
We shall also extend Theorem 3.3.1 to cover the cases of negative and fractional indices.

A summary of the rules of both this chapter and the last will be given at the end of the chapter.

5.2. The reciprocal of a function and quotient of two functions

Before we prove the general result for the reciprocal of a function, let us look at a special case [cf Examples 3.3.2, Number 6]:

If $g : \mathbb{R} - \{0\} \longrightarrow \mathbb{R}$ is defined by $g(x) = \frac{1}{x}$ then g is differentiable and $Dg(x) = \frac{-1}{x^2}$

To see this let x_0 be any point in $\mathbb{R} - \{0\}$, then

$$\frac{1}{x} = \frac{1}{x_0} + \left(\frac{1}{x} - \frac{1}{x_0}\right) \quad \text{for all } x \neq 0$$
$$= \frac{1}{x_0} + (x - x_0)\left(\frac{-1}{x_0 x}\right)$$

In other words
$$g(x) = g(x_0) + (x - x_0)q_{x_0}(x)$$

where
$$q_{x_0}(x) = \frac{-1}{x_0 x}$$

is continuous at x_0, as the reciprocal of a never zero continuous function is continuous.

So g is differentiable at x_0 and $Dg(x) = -1/x^2$

Now we can consider the following theorem:

Theorem 5.2.1. *If $f : A \longrightarrow \mathbb{R}$ is differentiable and non-zero for all $x \in A$ then*
$$\frac{1}{f} : A \longrightarrow \mathbb{R}$$
is differentiable and
$$D\left(\frac{1}{f}\right) = -\frac{Df}{f^2}$$

Proof. $\frac{1}{f}$ can be written as the composition of the reciprocal function $g(x) = \frac{1}{x}$ and the original function f:-
$$\left(\frac{1}{f}\right)(x) = (g \circ f)(x)$$

So by Theorem 4.3.1
$$D\left(\frac{1}{f}\right) = (Dg \circ f) \cdot Df$$
$$= \frac{-1}{f^2} \cdot Df \quad \text{by the example immediately before this theorem}$$
$$= -\frac{Df}{f^2}$$

$\square$

Using this result we can prove our next theorem.

Theorem 5.2.2. *If $f : A \longrightarrow \mathbb{R}$ is differentiable and $g : B \longrightarrow \mathbb{R}$ is differentiable and non-zero for all $x \in B$ then*
$$\frac{f}{g} : A \cap B \longrightarrow \mathbb{R}$$
is differentiable and
$$D\left(\frac{f}{g}\right) = \frac{Df \cdot g - f \cdot Dg}{g^2}$$

5.2. THE RECIPROCAL OF A FUNCTION AND QUOTIENT OF TWO FUNCTIONS

Proof. By the product and reciprocal results, (Theorems 4.2.2 and 5.2.1)

$$\frac{f}{g} = f \cdot \frac{1}{g}$$

is differentiable and

$$D\left(\frac{f}{g}\right) = D\left(f \cdot \frac{1}{g}\right)$$
$$= f \cdot D\left(\frac{1}{g}\right) + \frac{1}{g} \cdot Df$$
$$= f \cdot \left(\frac{-Dg}{g^2}\right) + \frac{Df}{g}$$
$$= \frac{-f \cdot Dg}{g^2} + \frac{g \cdot Df}{g^2}$$
$$= \frac{Df \cdot g - f \cdot Dg}{g^2}$$

$\square$

Examples 5.2.1

1. If

$$f(x) = \frac{x^2 - 1}{x^2 + 1}$$

then $f'(x) = \dfrac{(2x)(x^2 + 1) - (x^2 - 1)(2x)}{(x^2 + 1)^2}$

$$= \frac{4x}{(x^2 + 1)^2}$$

2. If

$$f(x) = \frac{x}{x^2 + 1}$$

then $f'(x) = \dfrac{(1)(x^2 + 1) - (x)(2x)}{(x^2 + 1)^2}$

$$= \frac{1 - x^2}{(x^2 + 1)^2}$$

3. If
$$f(x) = x^{-n} = \frac{1}{x^n}$$
$$\text{then} \quad f'(x) = -\frac{nx^{n-1}}{(x^n)^2}$$
$$= -\frac{nx^{n-1}}{x^{2n}}$$
$$= (-n)x^{-n-1}$$

So Theorem 3.3.1 holds for both positive and negative integers. If we interpret x^0 to be 1, then Theorem 3.3.1 is true for $n = 0$ as well, in other words true for all $n \in \mathbb{Z}$.

5.3. The inverse of a function

We now come to our last rule which is the rule for differentiating the inverse of a function.

Theorem 5.3.1. *If* $f : A \longrightarrow \mathbb{R}$ *is one to one and differentiable and f^{-1} is continuous and if $Df(x) \neq 0$ for all $x \in A$, then f^{-1} is differentiable and*
$$D(f^{-1}) = \frac{1}{Df \circ f^{-1}}$$

Proof. f is differentiable at $x_0 \in A$, so there exists a function $p_{x_0} : A \longrightarrow \mathbb{R}$, continuous at x_0, such that
$$f(x) = f(x_0) + (x - x_0)p_{x_0}(x) \tag{5.1}$$
Now let $y = f(x)$ so $x = f^{-1}(y)$, then equation 5.1 reads
$$y = y_0 + (f^{-1}(y) - f^{-1}(y_0))p_{x_0}(f^{-1}(y)) \tag{5.2}$$
Rearranging this equation we get
$$f^{-1}(y) = f^{-1}(y_0) + (y - y_0)\frac{1}{p_{x_0}(f^{-1}(y))} \tag{5.3}$$
Thus f^{-1} is differentiable at y_0 and hence on all of $f(A)$ and
$$Df^{-1}(y_0)) = \frac{1}{Df(f^{-1}(y_0))}$$
So
$$D(f^{-1}) = \frac{1}{Df \circ f^{-1}}$$
□

5.3. THE INVERSE OF A FUNCTION

Using this result we can extend Theorem 3.3.1 to fractional indices by saying:

$$\text{Let } f_n(x) = x^n \quad \text{for all } x \quad \text{if } n \text{ is odd}$$
$$\text{and } f_n(x) = x^n \quad \text{for } x \geq 0 \quad \text{if } n \text{ is even.}$$

Then f_n is a differentiable, one to one function whose inverse function is

$$g_n(x) = \sqrt[n]{x} = x^{\frac{1}{n}}$$

By Theorem 5.3.1 we have for $x \neq 0$

$$g'_n(x) = \frac{1}{f'_n(f_n^{-1}(x))}$$
$$= \frac{1}{n(f_n^{-1}(x))^{n-1}}$$
$$= \frac{1}{n\left(x^{\frac{1}{n}}\right)^{n-1}}$$
$$= \frac{1}{n} \cdot \frac{1}{x^{1-\left(\frac{1}{n}\right)}}$$
$$= \frac{1}{n} \cdot x^{\frac{1}{n}-1}$$

Now consider

$$f(x) = x^{\frac{m}{n}} = \left(x^{\frac{1}{n}}\right)^m$$

then by the Chain Rule

$$f'(x) = m\left(x^{\frac{1}{n}}\right)^{m-1} \cdot \frac{1}{n} \cdot x^{\frac{1}{n}-1}$$
$$= \frac{m}{n} \cdot x^{[(\frac{m}{n}-\frac{1}{n})+(\frac{1}{n}-1)]}$$
$$= \frac{m}{n} \cdot x^{\frac{m}{n}-1}$$

So now we can differentiate $f(x) = x^a$ for any rational number a. An examination of irrational exponents will have to be saved for later.

5.4. Summary of results from Chapters 4 and 5

$$\text{Sum rule} \quad D(f+g) = Df + Dg$$

$$\text{Product Rule} \quad D(f \cdot g) = f \cdot Dg + g \cdot Df$$

$$\text{Chain Rule} \quad D(f \circ g) = (Df \circ g) \cdot Dg$$

$$\text{Reciprocal rule} \quad D\left(\frac{1}{f}\right) = \frac{-Df}{f^2}$$

$$\text{Quotient rule} \quad D\left(\frac{f}{g}\right) = \frac{Df \cdot g - f \cdot Dg}{g^2}$$

$$\text{Inverse Function rule} \quad D(f^{-1}) = \frac{1}{Df \circ f^{-1}}$$

If $f(x) = x^a$, $x \in \mathbb{Q}$, then $f'(x) = ax^{a-1}$.

5.5. Exercises

1. Find Df for each of the following functions:-
(a)
$$f(x) = \frac{x^2 + x}{1 + x^2}$$
(b)
$$f(x) = \left(\frac{x + 2x^3}{2 + x^3}\right)^3$$
(c)
$$f(x) = \frac{x}{1+x}$$
(d)
$$f(x) = \frac{x}{1+|x|}$$

2. For each of the functions in Question 1, state exactly where that function is differentiable.

Chapter 6
Significance of the Derivative

6.1. Objectives

In this chapter we meet the ideas of maximum and minimum points of a function and that of critical points, drawing a distinction between them.

We will also prove two very important theorems: Rolle's theorem and the Mean Value Theorem.

6.2. New Terms introduced

maximum point	87
maximum value	87
minimum point	87
minimum value	87
local maximum (minimum) point	89
critical point	90
critical value	90

6.3. Critical Points and Critical Values

In both this section and the next we will be extracting information about a function from information about its derivative. We start by looking at maximum points.

Definition 6.3.1. *Let f be a function and $[a, b]$ a closed interval contained in the domain of f. A point $x \in [a, b]$ is a <u>maximum point</u> for f on $[a, b]$ if*
$$f(x) \geq f(y) \quad \text{for every } y \in [a, b]$$

The number $f(x)$ is called the <u>maximum value</u> of f on $[a, b]$.

Notice that although a function can have several different maximum points it has at most one maximum value.
We will leave the definition of minimum points and a minimum value for a function to you.
(You might use:- f has a minimum point on $[a,b]$ at x if $-f$ has a maximum point there, for instance).

We can now prove:-

Theorem 6.3.1. *Let f be any function defined on (a,b). If x is a maximum (or minimum) point for f on (a,b) and f is differentiable at x then $Df(x) = 0$.*

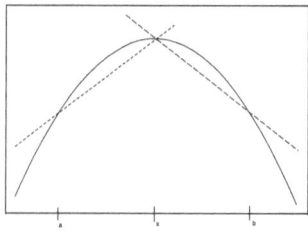

The idea of the proof is that chords drawn through points to the left of $(x, f(x))$ will have positive slope, while chords drawn through points to the right of $(x, f(x))$ have negative slope. So the slope of the tangent at $(x, f(x))$ must be zero.

Figure 6.1

Proof. If $h > 0$ and is a number such that $x + h \in (a,b)$ then as x is a maximum point we know that $f(x) \geq f(x+h)$. So we can deduce that $f(x+h) - f(x) \leq 0$ and letting p_x be the chord slope function at x we have
$$p_x(x+h) = \frac{f(x+h) - f(x)}{h} \leq 0$$
If $h < 0$ and is a number such that $x + h \in (a,b)$ then as x is a maximum point we know that $f(x) \geq f(x+h)$. So again we can deduce that
$f(x+h) - f(x) \leq 0$ and remembering that $h < 0$ we have
$$p_x(x+h) = \frac{f(x+h) - f(x)}{h} \geq 0$$
But f is differentiable at x and this means that p_x is continuous there, so the only possibility for $p_x(x)$ is that it is zero. hence
$$p_x(x) = 0 \text{ and so } Df(x) = 0$$

□

The proof for a minimum point is left to you, one line should do it!

If you were following the proof above closely you will have seen that we have used a result without actually declaring or proving it. That result is:-

Let $f : (a,b) \longrightarrow \mathbb{R}$ be continuous at $x_0 \in (a,b)$. If $f(x) \leq 0$ for all $x < x_0$ and $f(x) \geq 0$ for all $x > x_0$, then $f(x_0) = 0$.

This result, although it seems obvious, is by no means trivial.

We will now look at a slightly larger class of points.

Definition 6.3.2. *Let $f : [a,b] \longrightarrow \mathbb{R}$ be a function. A point $x \in [a,b]$ is a <u>local maximum (minimum) point</u> for f on $[a,b]$ if there is some $\delta > 0$ such that x is a maximum (minimum) point for f on $[a,b] \cap (x - \delta, x + \delta)$.*

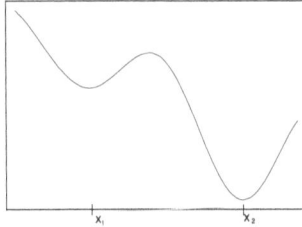

x_1 is a local minimum point
x_2 is a local maximum point

Figure 6.2

Applying Theorem 6.3.1 we can prove very easily (in fact it is left to you to do):

Theorem 6.3.2. *If $f : (a,b) \longrightarrow \mathbb{R}$ has a local maximum (or minimum) point at x and f is differentiable at x then $Df(x) = 0$.*

The converse of this theorem is not true, $Df(x) = 0$ does not mean that x is a local maximum or minimum point for f. For example if $f(x) = x^3$ then $Df(x) = 0$ when $x = 0$ but there is no local maximum or minimum for this function anywhere!

Thus we need a name for points where the derivative is zero.

Definition 6.3.3. *A critical point of a function f is a number x such that*
$$Df(x) = 0$$
The number $f(x)$ is called a *critical value* of f.

Now suppose we want to find the maximum or minimum value of f on a closed interval $[a, b]$.
We must look at three kinds of points

$$\text{The critical points of } f \text{ in } (a,b) \tag{6.1}$$
$$\text{The end points } a \text{ and } b \tag{6.2}$$
$$\text{Points where f is not differentiable.} \tag{6.3}$$

If x does not belong to class 6.2 or 6.3 then it belongs to (a,b) and f is differentiable there, so Theorem 6.3.1 applies.
So all one has to do is take all the points where f is not differentiable, all the critical points and the two end points, evaluating f at each one of them. The largest value will be the maximum value of f and the smallest, the minimum.

Examples 6.3.1
1. We shall find the maximum and minimum values of the function
$$f(x) = x^3 - x \text{ on } [-1, 2]$$

Now $Df(x) = 3x^2 - 1$
So $Df(x) = 0$ when $3x^2 - 1 = 0$ that is when $x = \pm\sqrt{1/3}$
The end points of the interval are -1 and 2 and f is differentiable everywhere.
So we can calculate:-

$$f\left(\sqrt{\frac{1}{3}}\right) = \frac{1}{3}\sqrt{\frac{1}{3}} - \sqrt{\frac{1}{3}} = -\frac{2}{3}\sqrt{\frac{1}{3}}$$
$$f\left(-\sqrt{\frac{1}{3}}\right) = -\frac{1}{3}\sqrt{\frac{1}{3}} + \sqrt{\frac{1}{3}} = \frac{2}{3}\sqrt{\frac{1}{3}}$$
$$f(-1) = 0$$
$$f(2) = 6$$

Thus the minimum value is $-\frac{2}{3}\sqrt{\frac{1}{3}}$, occurring at $x = \sqrt{\frac{1}{3}}$ and the maximum value is 6 occurring at $x = 2$

If we have to find maximum or minimum points in an open interval, where they may not exist, we use a little low cunning.

2. Suppose we are looking for the maximum and minimum, if they exist, of
$$f(x) = \frac{1}{1-x^2} \text{ on } (-1, 1).$$

Well $Df(x) = \frac{2x}{(1-x^2)^2}$, so $Df(x) = 0$ only when $x = 0$.

When x is close to 1 or -1, $f(x)$ becomes arbitrarily large so there is certainly no maximum value.
However we can show that f has a minimum at 0.

There are numbers a and b such that $-1 < a < 0$ and $0 < b < 1$ with $f(x) > f(0)$ for $x \in (-1, a]$ and for $x \in [b, 1)$.

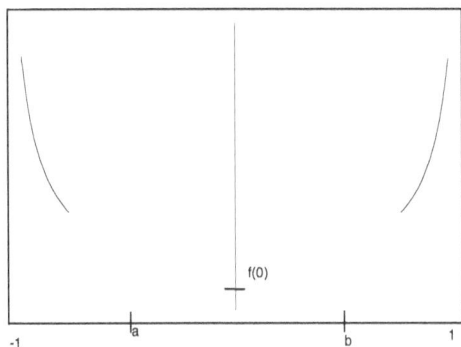

Figure 6.3

So the minimum of f on $[a, b]$ is the minimum of f on $(-1, 1)$. On $[a, b]$ the minimum occurs at 0, since a and b are discarded by definition.

6.4. The Mean Value Theorem

Now let us look at a simple question.
If $f'(x) = 0$ must f be a constant function?
Intuitively the answer should be yes.
Yet in order to actually prove this we need an important result known as the Mean Value Theorem.

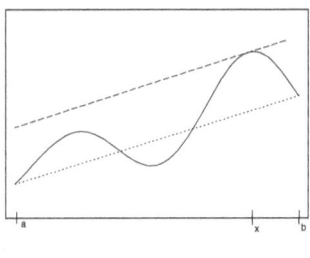

Figure 6.4

This says that if f is differentiable on $[a, b]$, then we can find a point on the graph of f such that the tangent line to that point is parallel to the line joining $(a, f(a))$ and $(b, f(b))$, in other words there exists $x \in (a, b)$ such that

$$Df(x) = \frac{f(b) - f(a)}{b - a}$$

We begin with a special case.

Theorem 6.4.1. *[Rolle's Theorem] If f is a continuous function on $[a, b]$ and differentiable on (a, b) and $f(a) = f(b)$ then there is an $x \in (a, b)$ such that $Df(x) = 0$.*

Proof. Since f is continuous on $[a, b]$ it must have a maximum and a minimum value on that interval.
Suppose first that the maximum value of f occurs at some point $x \in (a, b)$. Then $Df(x) = 0$ by Theorem 6.3.1 and the theorem is proved.
Suppose next that the minimum value of f occurs at a point $x \in (a, b)$. Then again by Theorem 6.3.1, $Df(x) = 0$ and we are finished.
Otherwise the maximum and minimum values occur at the end points a and b. Since $f(a) = f(b)$ the maximum and minimum values of f are the same, so f is a constant function and thus we can choose any $x \in (a, b)$ to satisfy the theorem.
□

We can now "tilt" Rolle's Theorem to get:-

6.4. THE MEAN VALUE THEOREM

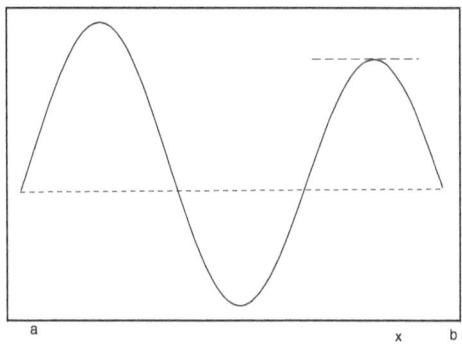

Figure 6.5: Rolle's Theorem

Theorem 6.4.2. *[The Mean Value Theorem] If f is a continuous function on $[a, b]$ and differentiable on (a, b) then there is an $x \in (a, b)$ such that*
$$Df(x) = \frac{f(b) - f(a)}{b - a}$$

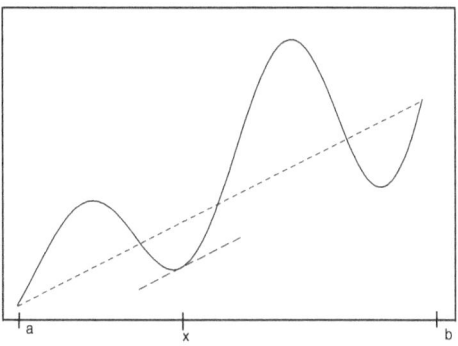

Figure 6.6: Tilted Rolle's Theorem = Mean Value Theorem

Proof. Define a new function g by
$$g(x) = f(x) + \frac{b - x}{b - a}[f(b) - f(a)]$$

g is continuous on $[a,b]$, since it is the sum of two continuous functions and is differentiable on (a,b), as it also the sum of two differentiable functions.
Furthermore $g(a) = f(b)$ and $g(b) = f(b)$.
Hence by Rolle's Theorem, there is an $x \in (a,b)$ such that $Dg(x) = 0$.
So
$$0 = Dg(x) = Df(x) + \frac{-1}{b-a}[f(b) - f(a)]$$
$$\text{that is } Df(x) = \frac{f(b) - f(a)}{b - a}$$

□

Now we can solve the problem from the beginning of this section:
Let a and b be any two points in the domain of f with $a \neq b$.

Then there is an $x \in (a,b)$ such that $Df(x) = \frac{f(b)-f(a)}{b-a}$.

But $Df(x) = 0$, for all x, so $f(b) = f(a)$.
Since a and b were any two points, f must be a constant function.

6.5. Exercises

1. Prove the result used in the section above, namely:-
Let $f : (a,b) \longrightarrow \mathbb{R}$ be continuous at $x_0 \in (a,b)$. If $f(x) \leq 0$ for all $x < x_0$ and $f(x) \geq 0$ for all $x > x_0$, then $f(x_0) = 0$.

2. Prove Theorem 6.3.2, stated in the section that:-
If $f : (a,b) \longrightarrow \mathbb{R}$ has a local maximum (or minimum) point at x and f is differentiable at x then $Df(x) = 0$.

3. For each of the following functions on the given closed intervals, find their maximum and minimum values:-
(a)
$$f(x) = x^3 - x^2 - 8x + 1 \text{ on } [-2, 2]$$
(b)
$$f(x) = \frac{x+1}{x^2+1} \text{ on } \left[-1, \frac{1}{2}\right]$$

(c)
$$f(x) = \frac{1}{x^5 + x + 1} \text{ on } \left[-\frac{1}{2}, 1\right]$$

4. Prove that the polynomial function $f_m(x) = x^3 - 3x + m$ never has two roots in $[0,1]$, no matter what the value of m. (Use Rolle's Theorem).

5. Prove that
$$\frac{1}{9} < \sqrt{66} - 8 < \frac{1}{8}$$
without computing $\sqrt{66}$ to 2 decimal places!!

Chapter 7
Averages

7.1. Objectives

This chapter introduces the ideas of the average of a function over a closed, or half closed interval. By the end of the chapter you should understand what the average of a function means and what properties it has. You should be able to calculate the average of simple functions over suitable intervals.

7.2. New Terms introduced

    Average value of the derivative, f',
    of a function over the closed interval I $M_I(f')$
    The Subdivision Property
    The Contraction Property
    Average value of a function $f : I \longrightarrow \mathbb{R}$ $M_I(f)$

7.3. Introduction

It is an accident of history that, because analytic geometry predates calculus, the ideas of differentiability have been expressed in terms of constructing a tangent to a curve and those of integrability in terms of finding an area under a curve. In order to make these two concepts precise, limits have had to be extensively used until the idea of a limit seem to dominate the entire field of analysis and to obscure the basic issues.

    But I intend to use a completely non-geometric definition of the integral which short-circuits the idea of limits.

    Why? Well much of the argument put forward to justify a similar decision in the case of continuity and differentiation applies here. In particular this definition makes many of the theorems that we have to prove, easier! Also it can be extended and generalized to situations where integration is needed but the concept of area is meaningless; for integration like differentiation, turns up in all sorts of unlikely places.

Thirdly, this new definition involves the idea of "anti-differentiation" in a natural way and thus, I hope, leads to a clearer understanding of why integration and anti-differentiation are connected than does the definition in terms of area, where it looks as if the mathematician is relying on sympathetic magic!

In this chapter I shall be considering the domain of any function, to be a closed interval $I = [a,b]$ for some a,b in the set $\mathbb{R}$.

7.4. Definition of an Average

Let I be an interval $[a,\ b]$ and $f : I \longrightarrow \mathbb{R}$ be a continuous function and also let $f : (a,b) \longrightarrow \mathbb{R}$ be differentiable, then the Mean Value Theorem tells us that there is some point c in (a,b) at which $f(b) - f(a) = (b-a)f'(c)$. In other words, we can find a point c in (a,b) such that

$$f'(c) = \frac{f(b) - f(a)}{b - a}$$

This value, $f'(c)$ is the mean or average value of the function f' over the interval (a,b), hence the name Mean Value Theorem. I am not going to be concerned with the precise point c at which f' has its average value, but with that average value itself.

There is also a technical almost nit-picking point, over whether the average can be defined on a closed interval,$[a,b]$ as well as an open interval, (a,b) or even defined over a half closed and half open interval, $[a,b)$ or $(a,b]$!!

The definition given above is quite clear that the average of f' is defined over the open interval (a,b). But suppose I want the average over the closed interval $[a,b]$? Because f is continuous on $[a,b]$, and even though f' is not defined at either of the end point a or b, what I want to say is that the average over the closed interval is the same as that over the open interval.

The continuity of f at a and b guarantee that the original function is well behaved at those points in terms of its values and so we can say that, (considering a first), although f is not differentiable at a, we could if we wanted, define a value for $Df(a)$ which would make Df continuous at a. Similarly we could do the same for the other end point, b.

Doing that means that it makes sense to talk about the average of f' over the closed interval $[a,b]$ and that this "new" average can be defined as numerically the same as the old without this leading to any problems..

7.4. DEFINITION OF AN AVERAGE

For that reason, I will denote the average value of the function f' over the interval (a, b), or the interval $[a, b]$, or the interval $[a, b)$ or the interval $(a, b]$ by the notation $M_I(f')$.

So, our formal definition goes:-

Definition 7.4.1. *Let I be an interval $[a, b]$ and $f : I \longrightarrow \mathbb{R}$ be a continuous function and also let $f : (a, b) \longrightarrow \mathbb{R}$ be differentiable, then the average value of the function f' over the interval I, $M_I(f')$, is defined as:-*

$$M_I(f') = \frac{f(b) - f(a)}{b - a}$$

Now the idea of "an average" is one that we all understand, fairly well, unless we have been got at by statisticians who have at least five different kinds of average! So if the number $M_I(f')$ is going to be an average, there are a couple of simple properties I would like it to have. The first of these is the Subdivision Property, which is illustrated in Figure 7.1.

What we do is split I into two subintervals $A = [a, t]$ and $B = (t, b]$, so $I = A \cup B$. Then we should expect that the average over I would be made up of the sum of the averages over A and B, each multiplied by a factor consisting of the proportion that subinterval is of the whole interval.

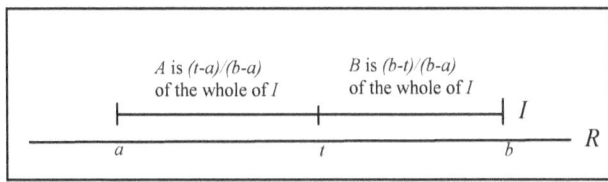

Figure 7.1: The Subdivision Property

Formally:-

Theorem 7.4.1. *[The Subdivision Property] Suppose $I = [a, b]$ is subdivided into two intervals $A = [a, t]$ and $B = (t, b]$, then*

$$M_I(f') = M_{A\cup B}(f') = \frac{t-a}{b-a}M_A(f') + \frac{b-t}{b-a}M_B(f')$$

Actually, having made so much fuss in stating it, the proof is terribly obvious.

Proof. Firstly,

$$M_A(f') = \frac{f(t)-f(a)}{t-a}, \text{ so } \frac{t-a}{b-a}M_A(f') = \frac{f(t)-f(a)}{b-a},$$

Secondly,

as $M_B(f') = \dfrac{f(b)-f(t)}{b-t}$, we will get $\dfrac{b-t}{b-a}M_B(f') = \dfrac{f(b)-f(t)}{b-a}$

Adding these two together gives us $M_I(f')$ again.

$\square$

My second required property, though slightly more subtle, is just as reasonable. This says that if we were to shrink the interval down to a point then the average of f' just becomes the value of f' at that point. We will call this property the Contraction Property.
Although it seems obvious, I must prove it. (Remember, in analysis, obvious does not mean true!)
So first let us state it formally.

Theorem 7.4.2. *[The Contraction Property] Let $f : [a,b] \longrightarrow \mathbb{R}$ be a continuous function with $f : (a,b) \longrightarrow \mathbb{R}$ being differentiable. Let x be any point in $[a,b]$. Construct a nested family of closed intervals: $[a,b] \supset [a_1,b_1] \supset [a_2,b_2] \supset [a_3,b_3] \supset \cdots \supset [a_n,b_n] \supset \ldots$ such that $x \in [a_n,b_n]$ for all values of n.*
Then, given an $\epsilon > 0$, we can find an $N \in \mathbb{Z}$ such that for all $n > N$,

$$\left|M_{[a_n,b_n]}(f') - f'(x)\right| < \epsilon$$

Proof. From our definition, using the Mean Value Theorem, we know that $M_I(f') = f'(c)$, for some c in (a,b).
Now imagine shrinking I to some point x in (a,b). If we stop the shrinking process at any stage we will still have a closed interval, $[u,v]$,

and the average of f' over $[u,v]$ will be equal to the value of f' at some point in the open interval (u,v).

Now continue the shrinking. We get a family of closed intervals whose interiors are neighbourhoods of x, each smaller closed interval contained in the all of previous closed intervals. (See Figure 7.2).

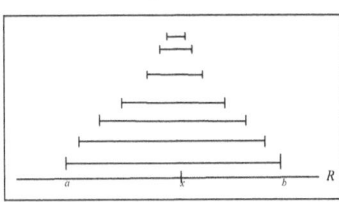

Figure 7.2: Shrinking family of closed intervals

Now on each of these closed intervals the average value of f' is the value of some f' in the neighbourhood of x, and so as f' is a continuous function, when the neighbourhoods are small the average value will be close to $f'(x)$.

How close is close? As close as we like, remember f' is continuous, hence it is bounded on a closed interval.

So the difference between $f'(x)$ and f' at some other point in the interior is bounded, and so by choosing a closed interval small enough we can make the difference between $f'(x)$ and the average value as close as we want.

□

So I have shown that the average value of f' over a closed interval has both the Subdivision and Contraction properties.

Actually I can do better than that. In fact, the average that I have defined is the only possible average having those two properties. Before I embark on showing that, let me say that this is a favourite game in all branches of mathematics. Having defined some mathematical entity, the definition is hotly pursued by "a uniqueness proof", that is, a demonstration that no other different entity can do just what the defined one does.

What is the point of a uniqueness proof? Well, besides setting one's mind at rest, so that when we talk about "the average value of f' " the word "the", as opposed to "a", is correct, it also means that if we later define an object, satisfying our two properties, for a bigger class of functions than just derivatives, we will know it agrees with our first definition for derivatives without having to do any extra work.

7.5. Uniqueness of an Average

How do I propose to prove that my average is unique? Simple, I shall assume that there is another such average and show that this leads me to an absurd result. (This is called a proof by contradiction.) First we state our result formally:-

Theorem 7.5.1. *Let $f : [a,b] \longrightarrow \mathbb{R}$ be a continuous function with $f : (a,b) \longrightarrow \mathbb{R}$ being differentiable, then the average value of the function f' over the interval I, $M_I(f')$, defined by:-*

$$M_I(f') = \frac{f(b) - f(a)}{b - a}$$

is the only function that satisfies both the Subdivision and the Contraction Properties.

Proof. Let $A_I(f')$ be a second "average' satisfying both the Subdivision and Contraction properties but being different from $M_I(f')$, that is, $A_I(f') \neq M_I(f')$.

Now it follows that either $A_I(f') > M_I(f')$ or $A_I(f') < M_I(f')$. We will assume the first possibility is the case. The proof for the second possibility will follow exactly similar lines and you should attempt it as an exercise.

So as $A_I(f') > M_I(f')$, we can write $A_I(f') = M_I(f') + d$ where d is a positive real number. Now we divide $[a,b]$ into two equal subintervals J and K, and, calculate our two different averages over each subinterval.

We will now show that either $A_J(f') = M_J(f') + d$ or $A_K(f') = M_K(f') + d$.

If this were not the case we would have the following situation:
$A_J(f') < M_J(f') + d$ and $A_K(f') < M_K(f') + d$.

Right, now use the Subdivision Property on $A_I(f')$ to say $A_I(f') = \frac{1}{2}[A_J(f') + A_K(f')]$.

However, the right hand side of this equation will be strictly less than
$\frac{1}{2}[M_J(f') + d + M_K(f') + d]$, using the inequalities above.
So $A_I(f') < \frac{1}{2}[M_J(f') + M_K(f') + 2d]$.
But $\frac{1}{2}[M_J(f') + M_K(f')] = M_I(f')$, by the Subdivision Property applied to $M_I(f')$, thus
$A_I(f') < M_I(f') + d$ which is $A_I(f')$ itself.
A number cannot be less than itself so we must have that either $A_J(f') = M_J(f') + d$ or $A_K(f') = MK(f') + d$ is true as required.

So we choose the subinterval for which the inequality holds and bisect that.

In one of these two new subintervals the "A" average will exceed the "M" average by at least d by the same argument again. So we continue our bisection of intervals and gradually we will contract down towards a point x.

Our "M" average can be made as close as we like to $f'(x)$, by the Contraction Property whilst our "A" average, which is at every stage d greater than the "M" average, must remain at least d greater than $f'(x)$. So the "A" average cannot be made as close as we like to $f'(x)$, thus A does not satisfy the Contraction Property. This shows, then, that our average is unique.

$\square$

Let us look at some examples.

Examples 7.5.1
1. If $f'(x) = c$ then to find the average value of f' over the interval $(0, 1)$ we need f itself.
   Well if $f(x) = cx$ then certainly $f'(x) = c$.
   So
   $$M_{[0,1]}(f') = \frac{f(1) - f(0)}{1 - 0} = \frac{c - 0}{1} = c$$

From the figure it is clear that the average value of f' must be c over any interval. As an exercise, let $I = [a,b]$ and show that $M_I(f') = c$.

You may have noticed that $f(x) = cx$ is not the only possible antiderivative for $f'(x) = c$. However, our uniqueness proof shows that a choice of a different anti-derivative could not give a different answer.

CHAPTER 7. AVERAGES

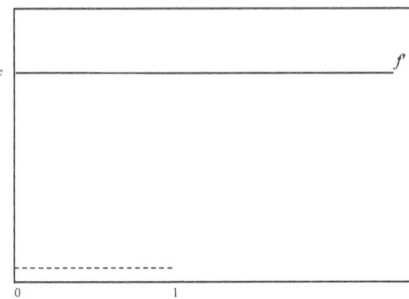

(Why?)

2. If $f'(x) = x$, then over the interval $[1,2]$, $f'(x)$ increases from 1 to 2 so we would expect its average value to be somewhere in between.

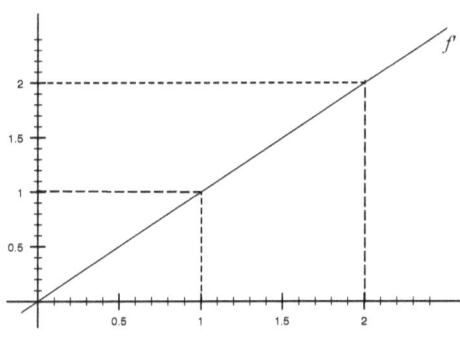

Let us calculate it.
Let
$$f(x) = \frac{x^2}{2}, \text{ then } f'(x) = x$$
So
$$M_{[1,2]}(f') = \frac{f(2) - f(1)}{2 - 1} = \frac{2 - \frac{1}{2}}{1} = \frac{3}{2}$$

3. If $f'(x) = x^2$, then we can find the average over the interval $[1, 3]$ by setting
$$f(x) = \frac{x^3}{3}$$

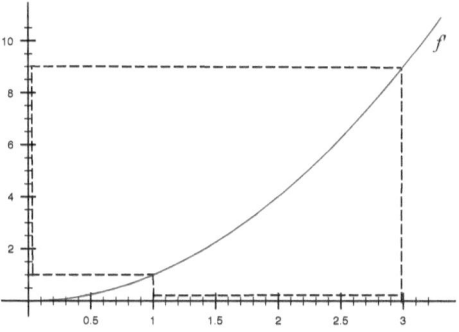

Then

$$M_{[1,3]}(f') = \frac{f(3) - f(1)}{3 - 1} = \frac{9 - \frac{1}{3}}{2} = \frac{13}{3}$$

7.6. What functions are Averageable?

We can now pause and ask a very important question. Which functions are averageable? Obviously those functions for which we can find an anti-derivative.

In fact we can say a little more than this. Any function which is piecewise anti-differentiable can be averaged. Piecewise anti-differentiable means that although over the whole interval there is no single anti-derivative f, yet if we subdivide the interval we can find an anti-derivative on each subinterval. We then calculate the average over each subinterval and use the Subdivision Property to find the average over the whole interval. Perhaps an example will help.

$$\text{Let } f'(x) = \begin{cases} 1, & \text{if } 0 \leq x < 1; \\ x + 1, & \text{if } 1 \leq x < 2; \\ x^2, & \text{if } 2 \leq x \leq 3. \end{cases}$$

So we can let

$$f(x) = \begin{cases} x, & \text{if } 0 \leq x < 1; \\ \frac{x^2}{2} + x, & \text{if } 1 \leq x < 2; \\ \frac{x^3}{3}, & \text{if } 2 \leq x \leq 3. \end{cases}$$

Then

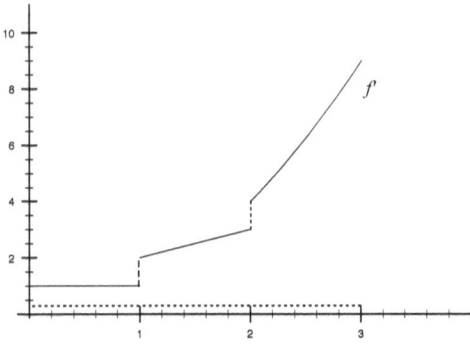

Figure 7.3

$$M_{[0,1]}(f') = \frac{f(1) - f(0)}{1 - 0} = \frac{1 - 0}{1} = 1$$

$$M_{[1,2]}(f') = \frac{f(2) - f(1)}{2 - 1} = \frac{4 - \frac{3}{2}}{1} = \frac{5}{2}$$

$$M_{[2,3]}(f') = \frac{f(3) - f(2)}{3 - 2} = \frac{9 - \frac{8}{3}}{1} = \frac{19}{3}$$

And

$$\begin{aligned}
M_{[0,3]}(f') &= \frac{1}{3}\left[M_{[0,1]}(f') + M_{[1,2]}(f') + M_{[2,3]}(f')\right] \\
&= \frac{1}{3}\left[1 + \frac{5}{2} + \frac{19}{3}\right] \\
&= \frac{1}{3} \times \frac{59}{6} \\
&= \frac{59}{18}.
\end{aligned}$$

However all we have done so far depends on our being able to find an anti-derivative for a function. In general this is by no means easy, and there are some very complicated, and some just downright sneaky ways of finding anti-derivatives. But faced with a function how are we going to know whether it has an anti-derivative or not? Oh, hadn't I said? Yes the horrible truth is that there are some innocent-looking functions which don't have anti-derivatives. For example, compare

7.6. WHAT FUNCTIONS ARE AVERAGEABLE?

$$h(x) = \sqrt{1 - x^2} \text{ and } g(x) = \sqrt{1 - x^4} \text{ for } -1 \leq x \leq 1$$

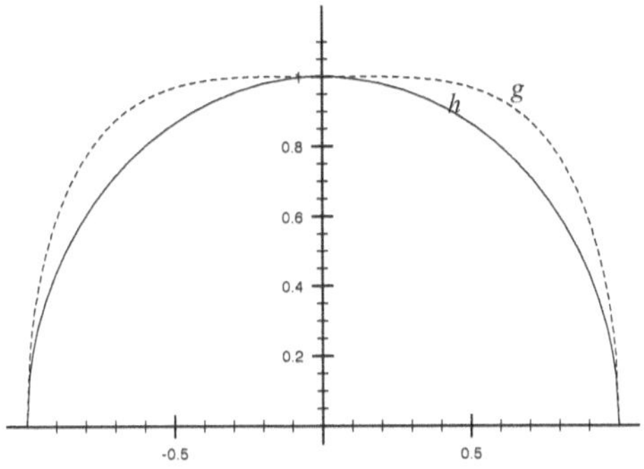

Figure 7.4

Using a sneaky method plus information about the sine function that we do not yet officially have, (in fact we don't yet officially know what a sine function is!!),I can anti-differentiate h and calculate that $M_{[-1,1]}(h) = \pi/4$, whatever π may be!

But no matter how sneaky I am, g just refuses to anti-differentiate, in fact it has no anti-derivative. Yet its graph is not that different from that of h and we would have expected it to have an average value close to that of h.

By using a simple numerical method, I can approximate $M_{[-1,1]}(g)$ as 0.87 which is close to $\pi/4$. Of course the simple numerical method depends on extending the definition of average to functions which cannot be anti-differentiated, so I am jumping the gun a bit in even mentioning it here!

So here is the problem. I have a function that I wish to average, $g(x)$, defined on a closed interval I for which I cannot find an anti-derivative. (Either because g has not got one, or just because I have not got enough low cunning to find it).

The obvious thing to try, is to use a function h which satisfies:

1. h is anti-differentiable and

2. h is a good approximation to g, by which I mean that for a given real number ϵ where $-\epsilon \leq g(x) - h(x) \leq \epsilon$, for all values of x in

I, I can show that $-\epsilon \leq M_I(g) - M_I(h) \leq \epsilon$. If I can do this then I have averaged my function g.

The programme then becomes to discover for what class of functions g, I can find suitable functions h.

First I must prove that if two functions are very close to each other on an interval then their average values on that interval are also close together.

And before I even do that, I must define the average value of a function that I cannot anti-differentiate.

Definition 7.6.1. *If $f : I \longrightarrow \mathbb{R}$ is a function defined on a closed interval I then, if it exists, the <u>average value of f over I</u>, is a number, denoted $M_I(f)$, chosen so that it satisfies both the Subdivision and the Contraction properties.*

We need to ask if this is a well-made definition. In other words, could there be two different choices of the average value of a function over an interval satisfying the two properties? No. Go back to the uniqueness proof earlier in this chapter. It shows that if you try to make a different choice then you *cannot* satisfy both properties.

But that was for the average value of a derivative, I hear you cry. Look at it carefully. Nowhere in the proof do I use the fact that f' is the derivative of f. Write it out again with f' replaced by f and you will see what I mean. So there is only one possible choice for the average of a function over an interval.

Furthermore, if f is anti-differentiable then we can use the formula given earlier to work out $M_I(f)$.

However, I still need a couple of elementary properties of the average, before proving the result about approximations.

Theorem 7.6.1. *The average of a non-negative function is non-negative, that is, if $f(x) \geq 0$ for all $x \in [a, b]$ then $M_{[a,b]}(f) \geq 0$.*

Proof. We prove this by the method of repeated bisection that we have used before.

Assume the result is false so $f(x) \geq 0$ on $[a, b]$ and $M_I(f) < 0$. Say $M_I(f) = -d$ where $d > 0$. Split I into two equal subintervals J and K.

Then by the Subdivision Property

$$M_I(f) = \frac{1}{2}[M_J(f) + M_K(f)]$$

and we will have $M_J(f) < 0$ or $M_K(f) < 0$ or both.
(If they were both greater than or equal to 0 then half their sum could not be negative).

Moreover either $M_J(f)$ or $M_K(f)$ must be less than or equal to $-d$. (Why?). Choose the interval on which the average is less than or equal to $-d$ and bisect again. By repeating this process we will get a contracting set of intervals on each of which the average of the function is less than or equal to $-d$. But these intervals are contracting to a point x and so we should be able to make the average value come as close as we like to $f(x)$. However, $f(x) \geq 0$ and yet the average value is always less than or equal to $-d$. So the contraction property is not satisfied, hence our result is proved. □

Theorem 7.6.2. *[The linearity property]*
If $M_{[a,b]}(g)$ and $M_{[a,b]}(h)$ exist and c and k are constants then

$$M_{[a,b]}(cg + kh) = c.M_{[a,b]}(g) + k.M_{[a,b]}(h)$$

This result is much easier to prove than it might appear.
In fact, after giving the broad outline of the proof, I am going to leave working out the fine details to you.
To prove this, you must show that the suggested average, that is the right hand side of the equation, satisfies the Subdivision and Contraction properties. The arguments will go like this:-

Firstly the Subdivision property.
Subdivide $[a,b]$ into $[a,t]$ and $[t,b]$.

Then $M_{[a,t]}(cg + kh)$, given by $cM_{[a,t]}(g) + kM_{[a,t]}(h)$ is the suggested average of $cg + kh$ over $[a, t)$.
Similarly for $[t, b]$.

Adding the two suggested averages together, each multiplied by its appropriate factor, ($\frac{t-a}{b-a}$ for $M_{[a,t]}(cg + kh)$) and using the fact that $M_I(g)$ and $M_I(h)$ obey the Subdivision property gives the result.

For the Contraction property, $M_I(g)$ contracts to $g(x)$ as I contracts towards x and $M_I(h)$ contracts to $h(x)$, so the right hand side will contract to $cg(x) + kh(x)$ as required.

We are now in a position to prove our Approximation Theorem.

Theorem 7.6.3. *If g and h are functions defined on a closed interval I, such that $M_I(g)$ and $M_I(h)$ exist and for a real number ε, $-\varepsilon \leq g(x) - h(x) \leq \varepsilon$, for all values of x in I, then $-\varepsilon \leq M_I(g) - M_I(h) \leq \varepsilon$.*

Proof. I will prove one of the two inequalities and leave the other to you as an exercise.

We are told that $g(x) - h(x) \leq \varepsilon$, so we can say that $\varepsilon + h(x) - g(x) \geq 0$ for all x in I. Now define a constant function on I by $E(x) = \varepsilon$, then we will have a function, $E + h - g$, defined on I which is greater than or equal to 0. So $M_I(E + h - g) \geq 0$ by Theorem 7.6.1.

Now using the linearity property we can say $M_I(E + h - g) = M_I(E) + M_I(h) - M_I(g)$.

In our example number 1 in Examples 7.5.1, $f(x) = c$, we found that $M_I(f) = c$ for any interval I, so $M_I(E) = \varepsilon$.

Hence we have $\varepsilon + M_I(h) - M_I(g) \geq 0$ so $M_I(g) - M_I(h) \leq \varepsilon$ as required.

The other inequality is proved by an exactly similar argument. □

Before going on any further I think we need a short break to look back and review our position.
So, the story so far.
We are given a function defined on a closed interval and we are trying to calculate its average value on that interval, where this average obeys two rules, that I have called the Subdivision and Contraction properties. We know that such an average, if it exists, is unique. We also know that if f is anti-differentiable then we can write down a specific formula for its average value over I.
We have proved some very important results about this average namely that, the average value of the constant function, $f(x) = c$, is c over any interval, the average of a non-negative function is non-negative and that the average obeys a linearity property.
Lastly we have proved that if two averageable functions are close together then so are their averages. What is left for us to do?
We still have to answer the question "which functions are averageable?" and that is quite a difficult question to answer.
The next chapter will seek to answer this question.

7.7. Exercises

1. Let $A_I(f')$ be a second "average' satisfying both the Subdivision and Contraction properties but being different from $M_I(f')$, that is, $A_I(f') \neq M_I(f')$.
   Prove that the assumption that $A_I(f') < M_I(f')$ leads to a contradiction.

2. Prove Theorem 7.6.2 [The Linearity Property] in full, namely:-

   If $M_{[a,b]}(g)$ and $M_{[a,b]}(h)$ exist and c and k are constants then
   $$M_{[a,b]}(cg + kh) = c.M_{[a,b]}(g) + k.M_{[a,b]}(h)$$

3. Prove the other inequality from Theorem 7.6.3, that is:-
   If g and h are functions defined on a closed interval I, such that $M_I(g)$ and $M_I(h)$ exist and for a real number ε, $-\varepsilon \leq g(x) - h(x)$, for all values of x in I, then $-\varepsilon \leq M_I(g) - M_I(h)$.

4. For each of the following functions, f, find their average over the indicated intervals:-
   (a)
   $$f(x) = 2x + 1 \text{ over } [0,2]$$
   (b)
   $$f(x) = \frac{1}{1+x^2} \text{ over } [1,2]$$
   Hint: cf Exercise 5.5 Question 1(c)
   (c)
   $$f(x) = x^3 + 3x^2 - 1 \text{ over } [-1,1]$$
   (d)
   $$f(x) = \begin{cases} x^2, & \text{if } 0 \leq x < 1; \\ 2x + 3, & \text{if } 1 \leq x < 2; \\ 3x^2, & \text{if } 2 \leq x \leq 3. \end{cases}$$

Chapter 8
Step Functions

8.1. Objectives

This chapter is concerned with proving that for any continuous function defined on a closed interval, an average exists over that interval. By the end of the chapter you should understand how a set of increasing lower step functions lead to a value for the average of a continuous function and also know how to construct such a set.

8.2. New Terms introduced

Step Function

Lower Step Function

8.3. Introduction

The whole of this chapter will be spent proving a fundamental theorem which is *The Existence Theorem for Averages*.

Theorem 8.3.1. *If $f : I \longrightarrow \mathbb{R}$ is a continuous function, I a closed interval, then $M_I(f)$ exists.*

In other words any continuous function on a closed interval is averageable.

In order to prove this we will be using a particular kind of function known as a step function. Let me draw a picture of a step function before formally defining it.

Like many concepts it is much easier to draw a picture than it is to explain in words. However, I shall do my best! Let $[a,b]$ be a closed interval and suppose that $x_1, \ldots, x_{n-1}$ are points in $[a,b]$ such that $a < x_1 < x_2 < \cdots < x_{n-1} < b$.

We can think of $[a,b]$ as being made up of subintervals $[a, x_1), [x_1, x_2), \ldots, [x_{n-2}, x_{n-1}), [x_{n-1}, b]$,

CHAPTER 8. STEP FUNCTIONS

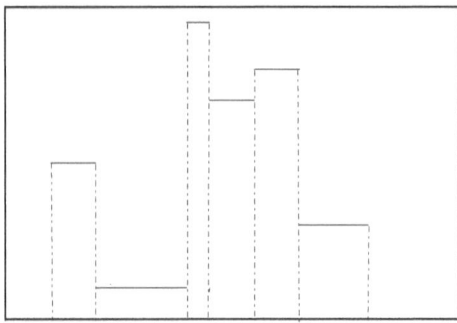

Figure 8.1: A step function

all of which, except for the last, are half open. Now a STEP FUNCTION defined on $[a,b]$ is a function which is constant on each of the subintervals but not necessarily constant on the whole interval.

Although our intuition tells us that a step function should be averageable and even how to calculate that average, there is a technical difficulty we must consider. Intuitively we might say, on each subinterval the function is constant so the average on that subinterval will be that constant value, then we use the Subdivision Property to calculate an average over $[a,b]$. The technical difficulty is, that except for the last one, all the subintervals are half open and we have only considered averages over closed intervals. (You may recall that in the last chapter we looked at averages for the derivative of a function over different sorts of intervals but not averages of a function in general. However the same basic approach used before will serve us well again here.)

We can get round this by making the most commonsense extension to our idea of average saying:

Definition 8.3.1. *Suppose f is a function defined and continuous on a half-open interval $[a,b)$. If we can define f(b) such that f is continuous on $[a,b]$ then we will define $M_{[a,b)}(f)$ by $M_{[a,b]}(f)$, if it exists.*

Notice that the condition that f should be continuous on $[a,b]$ rules out the possibility of there being two different choices for $f(b)$ (Prove this). Note too we can also consider averages over half-open intervals

$(a,b]$ in the obvious way, and hence also that we could define averages over open intervals, but we do not need to. Of course we ought to check that $M_{[a,b)}(f)$ satisfies the Subdivision and Contraction Properties. Can I leave that to you? Perhaps I ought to give you a hint. For the subdivision property just use the facts that the length of the interval $[a,b)$ is the same as that of the interval $[a,b]$ and that $M_{[a,b]}(f)$ satisfies the Subdivision Property. As for the Contraction Property, well, a point inside the interior of $[a,b)$ must be in the interior of $[a,b]$ as both interiors are the same, namely (a,b).

It is now obvious that a step function is averageable. But how do I intend to use them? Very simply. Given a function, f, that I want to prove averageable, I will construct a step function which will "lie below" the function. Then I will construct a set of step functions by tweaking the first step function to bring it closer to the original function while increasing its average, then doing the same for this second step function etc.etc. At the same time I will show that the set of step function averages that I construct is bounded above, then I will have constructed a possible average for f. Then all I need to do is to show that this definition satisfies the two properties of averages and then f is proved averageable.

I shall carry out this programme in three stages. Firstly I will show how to define the step function once we are given a continuous function, f. Secondly I will show how to define the average of f. Then finally, I will prove that this definition satisfies the Subdivision and Contraction Properties. I will then have proved that any continuous function is averageable!

8.4. Construction of the Lower Step Function

Suppose $f : I \longrightarrow \mathbb{R}$ is a continuous function defined on the closed interval $I = [a,b]$. Select points $x_1, .., x_{r-1}$ in $[a,b]$ so that $a < x_1 < x_2 < \cdots < x_{r-1} < b$. Since f is a continuous function it is bounded on every closed subinterval of $[a,b]$ and attains its bounds.

So we define our lower step function, h by saying, that for all x in $[x_{k-1}, x_k)$ we let $h(x)$ be the minimum value of f on $[x_{k-1}, x_k]$. For x in $[a, x_1)$ we let $h(x)$ be the minimum of f on $[a, x_1]$, and for x in $[x_{r-1}, b]$ we put $h(x)$ equal to the minimum of f on $[x_{r-1}, b]$. Clearly the step function I have defined depends on the choice of subdivision points in $[a,b]$. We can write down the value of $M_I(h)$ very easily. Just to make things look neater we will replace a by x_0 and b by x_r

CHAPTER 8. STEP FUNCTIONS

so our subdivision becomes:
$$a = x_0 < x_1 < \cdots < x_{r-1} < x_r = b$$

On each closed subinterval $[x_{k-1}, x_k)$ let m_k stand for the minimum value of f on that subinterval.
Then:

$$M_I(h) = \sum_{k=1}^{r} \frac{x_k - x_{k-1}}{b-a} m_k = \frac{1}{b-a} \sum_{k=1}^{r} m_k(x_k - x_{k-1})$$

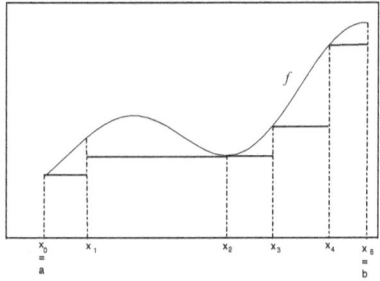

Figure 8.2: A step function

8.5. Definition of the Average of a Continuous Function over an interval

As f is continuous on $[a,b]$ it will be bounded as well. So suppose m is any lower bound for f on $[a,b]$ and M any upper bound. For each value of k, m_k must lie between m and M i.e. $m \leq m_k \leq M$.

So, $m(x_k - x_{k-1}) \leq m_k(x_k - x_{k-1}) \leq M(x_k - x_{k-1})$ as $x_k > x_{k-1}$. Furthermore, if we add up all these for every value of k we will get:

$$\sum_{k=1}^{r} m(x_k - x_{k-1}) \leq \sum_{k=1}^{r} m_k(x_k - x_{k-1}) \leq \sum_{k=1}^{r} M(x_k - x_{k-1})$$

However,

$$\sum_{k=1}^{r} m(x_k - x_{k-1}) = m \sum_{k=1}^{r} (x_k - x_{k-1})$$

because m is constant and so can be taken out of the summation as a common factor. The sum is then equal to $m(b - a)$. (Check this!)
So,

$$m(b-a) \leq \sum_{k=1}^{r} m_k (x_k - x_{k-1}) \leq M(b-a)$$

(because we can take M out of the sum $\sum_{k=1}^{r} M(x_k - x_{k-1})$ as a common factor, just as we did m for $\sum_{k=1}^{r} m(x_k - x_{k-1})$ and so the second sum is equal to $M(b - a)$).

Dividing by $(b - a)$, we get that $m \leq M_I(h) \leq M$.

So whatever the subdivision, the value of $M_I(h)$ is bounded above by M.

Let A be the set of all averages of step functions for f taken over all possible subdivisions of $[a,b]$. The set A is bounded above by M, thus by the completeness property it has a supremum. Now we define $M_{[a,b]}(f)$ to be the supremum of this set A. Putting all this formally we have:

Definition 8.5.1. *Suppose that $f : [a,b] \longrightarrow \mathbb{R}$ is continuous on $[a,b]$. For each subdivision of $[a,b]$ of the form*

$$a = x_0 < x_1 < \cdots < x_{r-1} < x_r = b$$

we can calculate the average of the lower step function for f (as shown above.)

Let $A \subset \mathbb{R}$ be the set of all the averages of step functions for f taken over all possible subdivisions of $[a,b]$. Then $M_{[a,b]}(f)$ is defined to be $\sup A$, which exists because A is bounded above.

8.6. Proof that this is an Average

This is the section where we will check that both the Subdivision and Contraction Properties hold.

First the Subdivision Property.
Let c be a point in $[a,b]$. Then construct a step function h_1 for f on $[a,c]$ and another, h_2, on $[c,b]$. This will necessitate constructing a subdivision both for $[a,c]$ and also for $[c,b]$. Putting them together will give a subdivision for $[a,b]$ and combining h_1 on $[a,c]$ with h_2 on

CHAPTER 8. STEP FUNCTIONS

$[c,b]$ will give a step function, h, for f on $[a,b]$ with

$$M_{[a,\,b]}(h) = \frac{c-a}{b-a} M_{[a,\,c]}(h_1) + \frac{b-c}{b-a} M_{[c,\,b]}(h_2) \quad (8.1)$$

Now if we are given any positive number, ϵ, we can find a subdivision of $[a,\,c]$ with corresponding step function h_1, so that:

$$M_{[a,\,c]}(h_1) > M_{[a,\,c]}(f) - \frac{\varepsilon(b-a)}{2(c-a)}$$

because $M_{[a,c]}(f)$ is the supremum of the set of averages of step functions for f on $[a,\,c]$.

Similarly we can find a subdivision of $[c,\,b]$ and step function h_2 defined on $[c,\,b]$ such that:

$$M_{[c,\,b]}(h_2) > M_{[c,b]}(f) - \frac{\varepsilon(b-a)}{2(b-c)}.$$

Combining these two inequalities with equation 8.1 above we have:

$$M_{[a,\,b]}(h) = \frac{c-a}{b-a} M_{[a,\,c]}(h_1) + \frac{b-c}{b-a} M_{[c,\,b]}(h_2)$$
$$> \frac{c-a}{b-a} M_{[a,\,c]}(f) + \frac{b-c}{b-a} M_{[c,\,b]}(f) - \varepsilon.$$

Because $M_{[\alpha,\beta]}(f)$ is the supremum of the set of step function averages for f over the closed interval $[\alpha,\beta]$, we can deduce:

$$M_{[a,\,b]}(f) \geq \frac{c-a}{b-a} M_{[a,\,c]}(f) + \frac{b-c}{b-a} M_{[c,\,b]}(f) - \varepsilon$$

and since this is the case for all positive ϵ, we can conclude that:-

$$M_{[a,\,b]}(f) \geq \frac{c-a}{b-a} M_{[a,\,c]}(f) + \frac{b-c}{b-a} M_{[c,\,b]}(f).$$

I will now prove that the inequality holds the other way round, so we must have equality.

Take a subdivision of $[a,b]$. By including c, if it is not already one of the subdivision points, we get a subdivision of $[a,c]$ and of $[c,b]$. If we start with a step function h defined for the original subdivision, we will now get a slightly different step function, H, by including c as a subdivision point.

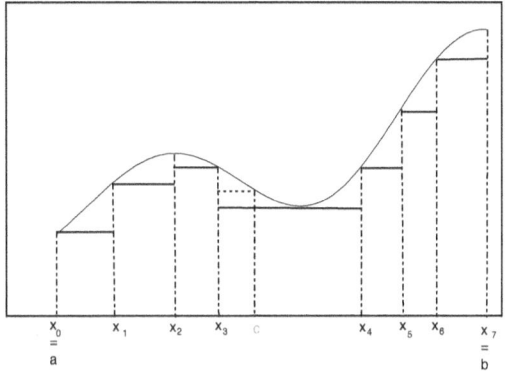

Figure 8.3: Including c into the partition

What will be the relation between $M_{[a,b]}(h)$ and $M_{[a,b]}(H)$?

A look at the figure should convince you that

$$M_{[a,b]}(H) \geq M_{[a,b]}(h)$$

Now, as an exercise, prove it!
Thus:

$$\frac{c-a}{b-a}M_{[a,\ c]}(f) + \frac{b-c}{b-a}M_{[c,b]}(f) \geq \frac{c-a}{b-a}M_{[a,\ c]}(h_1) + \frac{b-c}{b-a}M_{[c,b]}(h_2) \quad (8.2)$$

$$= M_{[a,b]}(H)$$
$$\geq M_{[a,\ b]}(h)$$

So the left hand side of inequality 8.2 is an upper bound for the set of step function averages, hence is greater than or equal to the supremum of that set namely $M_{[a,b]}(f)$.

In symbols,

$$\frac{c-a}{b-a}M_{[a,\ c]}(f) + \frac{b-c}{b-a}M_{[c,b]}(f) \geq M_{[a,\ b]}(f)$$

We have now shown the inequality holds both ways round, thus the only possibility is that both sides of the inequality are equal to

CHAPTER 8. STEP FUNCTIONS

each other, and this means that the Subdivision property holds.

Now for the Contraction Property.
I will give an outline of the proof and leave it to you to work it out in its full detail.

By contracting our interval towards a point x we can make $M_I(h)$ as close as we like to $h(x)$ and furthermore by choosing a large enough number of intervals in our subdivision we can make $M_I(h)$ as close as we like to $M_I(f)$. But as the number of subintervals increases, $h(x)$ and $f(x)$ will be getting closer to each other. Putting all this together will prove that the Contraction Property holds.

We have now shown that every continuous function defined on a closed interval is averageable over that interval. In the next chapter we will reinterpret our results using the idea of "the integral" and then relate that concept to differentiation.

8.7. Exercises

1. Take a subdivision of $[a,b]$. Include c, if it is not already one of the subdivision points, to get a subdivision of $[a,c]$ and of $[c,b]$. Consider a step function h defined for the original subdivision, by including c as a subdivision point, we will now get a slightly different step function, H. Prove that:-
$$M_{[a,b]}(H) \geq M_{[a,b]}(h)$$

2. Prove that $M_{[a,b]}(f)$, as defined in Definition 8.5.1 possesses the Contraction Property.

3. For the function $g(x) = \sqrt{1-x^4}$, defined on the closed interval [-1, 1], construct a lower step function on a partition of 6 equal width subintervals and calculate the average for that step function. Now double the number of subintervals in the partition, in other words bisect each partition, and again calculate the average of the step function.

Chapter 9
The Integral of a Function

9.1. Objectives

This chapter restates the results from the previous one in terms of the definite integral. By the end of the chapter you will know what is meant by the definite integral of a function over an interval, the elementary properties of that integral and how the process of integration is related to differentiation. You will also have met the two most common "methods" of integration, integration by substitution and by parts.

9.2. New Terms introduced

Definite Integral of f over $[a, b]$ $\qquad \int_{[a,b]} f$

First Fundamental Theorem of Calculus

Second Fundamental Theorem of Calculus

Integration by Substitution

Integration by Parts

9.3. Introduction

In the previous two chapters I have proved all the important properties of "integration" using the idea of the average of a function. Now I must introduce the idea of the integral and restate all the results.

Suppose that $f : I \longrightarrow \mathbb{R}$ is a continuous function. Then by the Existence Theorem, Theorem 8.3.1, $M_{[a.b]}(f)$ exists. By, <u>the definite integral of f over $[a, b]$</u> I will mean the number

$$(b - a) \times M_{[a,b]}(f)$$

This number will be written as:-

$$\int_{[a,b]} f$$

9.4. Elementary properties of the definite integral

All of these will follow immediately from the corresponding properties for averages.

The integral of a function satisfies the linearity property, so that

$$\int_{[a,b]} (cf + kg) = c \int_{[a,b]} f + k \int_{[a,b]} g$$

where c and k are real numbers and f and g continuous functions defined on $[a,b]$.
This property follows from the linearity property of averages, Theorem 7.6.2 by just multiplying right through by $(b - a)$.

The integral of a function satisfies the additive property, that is

$$\int_{[a,b]} f + \int_{[b,c]} f = \int_{[a,c]} f$$

To prove this all I need to do is to apply the subdivision property to the interval $[a,c]$ which will give me

$$M_{[a,c]}(f) = \frac{b-a}{c-a} M_{[a,b]}(f) + \frac{c-b}{c-a} M_{[b,c]}(f)$$

and then multiply through by $(c - a)$.

The non-negative property of the integral which says that:

If $f \geq g$ on $[a,b]$ then $\int_{[a,b]} f \geq \int_{[a,b]} g$.

The proof of this goes like this.
If $f \geq g$ on $[a,b]$ then $f - g \geq 0$ on $[a,b]$. As the average of a non-negative function is non-negative, we know that $M_{[a,b]}(f - g) \geq 0$. Multiplying by $(b-a)$ tells us that $\int_{[a,b]} (f - g) \geq 0$. Then the linearity property completes the proof.

The integral of a function over a point is zero that is $\int_{[a,a]} f = 0$.

Since $M_{[a,a]}(f) = f(a)$, multiplying by $(a - a) = 0$ gives the result.

The integral over [b,a] is minus that over [a,b], in symbols,

$$\int_{[a,b]} f = -\int_{[b,a]} f$$

To prove this note that $\int_{[a,b]} f + \int_{[b,a]} f$ by the additivity property, gives us $\int_{[a,a]} f$, which is zero by the previous result. Our property is now proved.

However, notice that for averages, $M_{[a,b]}(f)$ is equal to $M_{[b,a]}(f)$, no minus sign appears!

9.5. The relationship between differentiation and integration

The reason for introducing the concept of the definite integral instead of sticking with the average is that there is a very easy and neat relationship between the integral and the derivative.

To explore this I will state and prove the two Fundamental Theorems of Calculus. Before that let me introduce a new word. I will say a function is INTEGRABLE over $[a,b]$ if the integral of that function over that interval exists, (in other words if the function is averageable over that interval).

First Fundamental Theorem of Calculus.
If $F : [a,b] \longrightarrow \mathbb{R}$ is a function which is differentiable and $DF : [a,b] \longrightarrow \mathbb{R}$ is an integrable function then

$$\int_{[a,b]} DF = F(b) - F(a).$$

This "theorem" is just a restatement in terms of integrals, of the very first definition of the average of the derivative of a function, Definition 7.4.1!

Second Fundamental Theorem of Calculus
Suppose f is an integrable function defined on $[a,b]$. Let $F : [a,b] \longrightarrow \mathbb{R}$ be defined by $F(t) = \int_{[a,t]} f$

Then F is continuous and differentiable on (a,b) and $DF = f$.
To prove this result all we need to show is that F is differentiable and $DF = f$, as differentiability implies continuity. So we choose a point s in (a,b) and construct the chord slope function $\rho_s(t)$
Now,

$$\rho_s(t) = \frac{F(t) - F(s)}{t - s}, \text{ provided } s \neq t, \text{ so } \rho_s(t) = \frac{1}{t-s}\int_{[a,t]} f - \int_{[a,s]} f$$

So, using the fact that the integral over $[a,t]$ is minus one times that over $[t,a]$, we can say that

For $s \neq t$,

$$\rho_s(t) = \frac{1}{t-s}\left(-\int_{[t,a]} f - \int_{[a,s]} f\right) = \frac{-1}{t-s}\left(\int_{[t,a]} f + \int_{[a,s]} f\right)$$

Which by the additive property of the integral becomes:-

$$\frac{-1}{t-s}\int_{[t,s]} f$$

Now again using the result that the integral over $[t,s]$ is minus one times that over $[s,t]$, we have that

$$\rho_s(t) = \frac{1}{t-s}\int_{[s,t]} f = M_{[s,t]}(f)$$

by the definition of the integral.

In order to prove F is differentiable, all we have to do now is to show that there is a choice for $\rho_s(s)$ which ensures that ρ_s is continuous at s.

However from the contraction property for averages we can say that for t close to s, $M_{[s,t]}(f)$ is close to $f(s)$. Also $M_{[s,s]}(f)$, which is $\rho_s(s)$, is $f(s)$. So, ρ_s will be continuous at s if we define $\rho_s(s)$ to be $f(s)$. This argument needs to be expressed rigorously in terms of the definition of continuity at a point, and that I leave to you as an exercise.

9.6. Further properties of the integral

We have now shown that for certain kinds of functions differentiation and integration can be regarded as inverse operations to each other. From this simple fact stem all the well-known methods of finding integrals that occur in an elementary calculus course. It is not my intention, nor is it part of Analysis, to deal with those. However, I am going to prove two more theorems, which involve integration and differentiation and which happen to form the basis of two of the most important methods of integration, because they are important theorems in Analysis in their own right.

9.6. FURTHER PROPERTIES OF THE INTEGRAL

Theorem 9.6.1. *Integration by substitution.*
Let $g : [a,b] \longrightarrow \mathbb{R}$ be differentiable on (a,b) and $f : [g(a), g(b)] \longrightarrow \mathbb{R}$ be a function that is differentiable on $(g(a), g(b))$, then

$$\int_{[g(a),g(b)]} f = \int_{[a,b]} (f \circ g).Dg \qquad (9.1)$$

Proof. To prove this notice first, that as f is differentiable on $(g(a), g(b))$ it is continuous there and hence integrable over $[g(a), g(b)]$.

Suppose then that we define a function $F : [g(a), g(b)] \longrightarrow \mathbb{R}$ by

$$F(x) = \int_{[g(a),x]} f$$

then the left hand side of 9.1 is just $F(g(b))$.

Thus as

$$\int_{[g(a),g(b)]} f = F(g(b))$$

then

$$F(g(a)) = 0, \text{ and } DF = f.$$

Now consider $D(F \circ g)$.
By the chain rule we can say that

$$D(F \circ g) = (DF \circ g).Dg = (f \circ g).Dg.$$

So by the first fundamental theorem of calculus,

$$\int_{[a,b]} (f \circ g).Dg = (F \circ g)(b) - (F \circ g)(a) = F(g(b)) - F(g(a)) = F(g(b))$$

Thus the theorem is proved.

$\square$

Now the second theorem.

CHAPTER 9. THE INTEGRAL OF A FUNCTION

Theorem 9.6.2. *Integration by Parts.*
If f and g are two functions defined and differentiable on [a,b] then

$$\int_{[a,b]} f.Dg = [(f.g)(b) - (f.g)(a)] - \int_{[a,b]} g.Df$$

Proof. This result follows from the product rule for differentiation in the following way:
The product rule says that $D(f.g) = f.Dg + g.Df$
Hence $f.Dg = D(f.g) - g.Df$
Since the function on the left hand side of the equation is equal to that on the right hand side their integrals over $[a,b]$ will also be equal, so:

$$\int_{[a,b]} f.Dg = \int_{[a,b]} (D(f.g) - g.Df) = \int_{[a,b]} D(f.g) - \int_{[a,b]} g.Df$$

by the linearity property of the integral.
The first fundamental theorem of calculus now completes the proof by telling us that

$$\int_{[a,b]} D(f.g) = (f.g)(b) - (f.g)(a)$$

□

This now concludes the theorems on integration. But before I embark on the next topic, I will just say that the integral as defined above, can be applied to calculating a whole range of physical properties such as areas, volumes of solids, lengths of curves, surface area of solids, moments, both in mechanics, giving centroids and moments of inertia, and in statistics, giving means and variances. The concept can be generalized to integrating over an area or a volume or extended to include functions which are discontinuous. The three chapters above only form the tip of the iceberg, a brief introduction to the field of integration.

9.7. Exercises

1. Prove rigorously the statement made at the end of the proof of the Second Fundamental Theorem of Calculus, namely:-

Define a function $\rho_s(t)$ by:

$$\rho_s(t) = \frac{F(t) - F(s)}{t - s}, \text{ provided } s \neq t$$

where $F : [a, b] \longrightarrow \mathbb{R}$ is defined by $F(t) = \int_{[a,t]} f$.

If we define $\rho_s(s)$ to be $f(s)$, then ρ_s will be continuous at s.

2. For each of the following functions (taken from Exercise 7.7, Question 4), f, find their integral over the indicated intervals:-
(a)
$$f(x) = 2x + 1 \text{ over } [0, 2]$$

(b)
$$f(x) = \frac{1}{1 + x^2} \text{ over } [1, 2]$$

(c)
$$f(x) = x^3 + 3x^2 - 1 \text{ over } [-1, 1]$$

(d)
$$f(x) = \begin{cases} x^2, & \text{if } 0 \leq x < 1; \\ 2x + 3, & \text{if } 1 \leq x < 2; \\ 3x^2, & \text{if } 2 \leq x \leq 3. \end{cases}$$

Next some important if strange results

3. Consider the function f_2, defined on $[0,1]$ by:-

$$f_2(x) = \begin{cases} 1, & \text{if } x = \{0, \frac{1}{2}, 1\}. \\ 0, & \text{for all other values of } x; \end{cases}$$

Show that
$$\int_{[0,1]} f_2 = 0$$

Hint: Construct the lower step function using the intervals $[0, \frac{1}{2})$ and $[\frac{1}{2}, 1]$. There is only one possible step function!

127

CHAPTER 9. THE INTEGRAL OF A FUNCTION

4. Consider the function f_3, defined on $[0,1]$ by:-

$$f_3(x) = \begin{cases} 1, & \text{if } x = \{0, \frac{1}{3}, \frac{1}{2}, 1\}. \\ 0, & \text{for all other values of } x; \end{cases}$$

Show that
$$\int_{[0,1]} f_3 = 0$$

5. Consider the function f_4, defined on $[0,1]$ by:-

$$f_4(x) = \begin{cases} 1, & \text{if } x = \{0, \frac{1}{4}, \frac{1}{3}, \frac{1}{2}, 1\}. \\ 0, & \text{for all other values of } x; \end{cases}$$

Show that
$$\int_{[0,1]} f_4 = 0$$

Now the general case

6. Consider the function f_N, defined on $[0,1]$ by:-

$$f_N(x) = \begin{cases} 1, & \text{if } x = \{0, \frac{1}{N}, \ldots, \frac{1}{4}, \frac{1}{3}, \frac{1}{2}, 1\} \text{ where } N \text{ is any positive integer.} \\ 0, & \text{for all other values of } x; \end{cases}$$

Show that
$$\int_{[0,1]} f_N = 0$$

And now for the puzzle

7. Consider the function f, defined on $[0,1]$ by:-

$$f(x) = \begin{cases} 1, & \text{if } x = \{0, \ldots, \frac{1}{n}, \ldots, \frac{1}{4}, \frac{1}{3}, \frac{1}{2}, \ldots, 1\} \text{ for all } n \text{ in the set of positive integers.} \\ 0, & \text{for all other values of } x; \end{cases}$$

Show that
$$\int_{[0,1]} f$$

can not be defined!
Hint: Show that it is impossible to construct a lower step function for f.

Clearly as N get larger and larger f_N is getting closer and closer to f, yet although f_N always has an integral and the value of that integral is always zero, f cannot be integrated. This is a major drawback with our definition of the integral. However other definitions of an integral, such as the Lebesgue integral, are possible which overcome this problem

Chapter 10
Sequences and Series

10.1. Objectives

By the end of this chapter you should know what is meant by a sequence and a series and the difference between them. You will understand the terms convergent and divergent sequences and series. You will be able to calculate the sum of a series when that exists. Lastly you will be able to apply a simple test to determine whether or not such a sum exists.

10.2. New Terms introduced

sequence	$\{a_n\}$
nth term of a sequence	a_n
limit	$\lim_{n \to \infty} a_n$
convergent sequence	
divergent sequence	
series	$\sum a_n$
sum of a series	$\sum_{n=1}^{\infty} a_n$
Ratio Test	
D'Alembert's Ratio Test	

10.3. Introduction

We now come to the fourth major concept in Analysis. That is the idea of a limit. I have avoided using limits in defining continuity or differentiability or integrability in order to show that limits are not needed for these definitions. My contention is that limits are actually harder to understand than these three other ideas. But now comes the time when limits are indispensable to our work!

10.4. Sequences and Limits

First we must look at *sequences*.

A sequence is just a list of real numbers separated by commas like 1, $\frac{1}{2}, \frac{1}{4}, \frac{1}{8}, \frac{1}{16}, \ldots$ or 3,7,9,11,13, ...

That is not a very mathematical definition so we will try and formalize it a little.

<u>First attempt</u> . A sequence is a collection of numbers in some definite order, the first one being called the first term, the second the second term and so on.

By labelling the numbers in the sequence, the first term, the second term etc. we are establishing a correspondence between the natural numbers 1, 2, 3, 4, ... and the terms in our sequence. Such a correspondence is called a MAPPING. If you have done some set theory the word will be familiar to you. But do not worry if it is not, it just means correspondence, as far as we are concerned. We can now make a Formal Definition.

Definition 10.4.1. *A SEQUENCE is a mapping between the set of natural numbers and the set of real numbers, so that for each natural number, n say, the real number associated to n is called the nth TERM and often written as* a_n.

Just remember that all we are trying to say is that a sequence is a list of real numbers.

It is sometimes useful to have a notation for a sequence and we use the nth term enclosed in curly parentheses to represent the whole sequence.
So $\{a_n\} = a_1, a_2, a_3, a_4, \ldots$

We are going to be interested in a particular kind of sequence, the sort that approaches a *limit*. Roughly speaking, suppose as we take our natural number n larger and larger, the corresponding terms in the sequence, a_n, are getting closer and closer to some fixed real number, a, then a is called the limit of the sequence.

For instance, consider 1, $\frac{3}{2}, \frac{1}{2}, \frac{3}{4}, \frac{1}{4}, \frac{3}{8}, \frac{1}{8}, \frac{3}{16}, \frac{1}{16}, \ldots$
The terms are getting smaller and smaller even allowing for the step

up after each step down, so we would think that this sequence is approaching zero.

On the other hand the sequence $1, \frac{1}{2}, \frac{3}{4}, \frac{1}{4}, \frac{7}{8}, \frac{1}{8}, \frac{15}{16}, \frac{1}{16}, \ldots$ is not approaching zero, since no matter how far along the sequence you go some terms are still close to 1.

Now let us attempt to put this idea into a definition.

Initial Definition "The sequence whose nth term is a_n approaches the LIMIT a", means that, we can take any open interval around a and eventually (that is after some particular value of n) all the remaining terms of that sequence will lie in that open interval.

In case you do not see how this definition does the trick, the strength lies in that phrase "any open interval". We can make that interval as small as we like and still, eventually, all the terms of the sequence will lie there. It may only be after the millionth term that they all lie in our interval but it will happen. Let us sharpen our definition to:

Second Definition "The sequence whose nth term is a_n, approaches the limit a", means that taking any open interval around a, there is some positive integer N with the property that a_N and all subsequent a_n's lie in that interval.

Put in terms of neighbourhood's we have :

Definition 10.4.2. *The sequence whose nth term is a_n approaches a limit a if, given any neighbourhood of a, there is some positive integer N with the property that a_N and all subsequent a_n's ($n > N$) lie in that neighbourhood. a is called the limit of the sequence (sometimes, the limit of the sequence as n tends to infinity) and a common notation used for expressing this is* $\lim a_n = a$ *or* $\lim_{n \to \infty} a_n = a$.

The theory of sequences and limits does not stop here. There is much, much more that we could do with these ideas. However, as we have a limited space we must stop here because we have developed all that we are going to need. Just let me introduce one more word that we shall be using. A sequence which approaches a limit is called CONVERGENT. If it does not approach a limit it is called divergent.

10.5. Series

The words "sequence" and "series" tend to be used interchangeably in a lot of people's minds. But in mathematics they mean different things. A sequence is a list of numbers with commas in between, a

CHAPTER 10. SEQUENCES AND SERIES

series is the result of adding up all those numbers. Now in the last section we looked at a few sequences that "went on for ever", in other words they were infinite sequences. So, this gives us a problem. We all know how to add a finite number of numbers together, but how do you add an infinite number of numbers together? Even doing it quickly you would never get to the end of the list! So we need some reasonable way of defining "infinite addition".

Suppose we have a sequence $a_1, a_2, a_3, a_4, a_5, \ldots$ Consider a second sequence, formed from the first by the following means:

$$A_1 = a_1$$

$$A_2 = a_1 + a_2$$

$$A_3 = a_1 + a_2 + a_3$$

$$A_4 = a_1 + a_2 + a - 3 + a_4$$

and so on.

So A_{100} is the sum of the first hundred terms of our original sequence. Now, surely, if this second sequence converged then its limit would be the obvious choice for our "infinite sum".

At this stage an example might help.

Take the sequence $1, \frac{1}{2}, \frac{1}{4}, \frac{1}{8}, \frac{1}{16}, \ldots$

Then $A_1 = 1$, $A_2 = 1\frac{1}{2}, A_3 = 1\frac{3}{4}, A_4 = 1\frac{7}{8}, A_5 = 1\frac{15}{16}, \ldots$ .and it is easy to see that

$$A_n = 2 - \frac{1}{2^{n-1}}$$

I will leave that to you to prove!

So as n increases does A_n approach 2? Yes. To prove it, take any neighbourhood of 2. Inside that neighbourhood there will be an open interval of the form $(2 - \varepsilon, 2 + \varepsilon)$ where ε is a small positive real. As long as $\frac{1}{2^{n-1}}$ is smaller than ε, A_n must lie in that open interval. A little bit of simple algebra will enable you to compute the condition on n for this to happen.

Now let us make the (inevitable) formal definition.

10.5. SERIES

Definition 10.5.1. *Suppose we have a sequence whose nth term is* a_n.
The SERIES, $\sum a_n$, *is defined as the sequence whose nth term,* A_n, *is given by*

$$A_n = a_1 + a_2 + a_3 + \cdots + a_n$$

If this second sequence is convergent then the series is called convergent and the number $\lim A_n$ *is called the SUM OF THE SERIES and denoted :*

$$\sum_{n=1}^{\infty} a_n$$

In fact A_N is very often denoted by $\sum_{n=1}^{N} a_n$ as this notation is very handy.

It can be used to pick out middle terms in a series, for instance:

$$\sum_{n=r}^{r+k} a_n \text{ means } a_r + a_{r+1} + a_{r+2} + \cdots + a_{r+k}$$

Again there is a great deal that we could do to investigate series. For us, however, it will be sufficient to find one condition that implies that a series converges and leave it at that.

But first another example.

Consider the series $1 + t + t^2 + t^3 + \ldots$ which we could write as $\sum_{n=0}^{\infty} t^n$. Assume that $|t| < 1$.

We know, from elementary algebra, the formula for the sum of a geometric series, and it so happens that the first k terms of this series is just that!

$$\text{So } \sum_{n=0}^{k} t^n = \frac{1 - t^k}{1 - t}$$

We can write this as

$$\frac{1}{1-t} - \frac{t^k}{1-t}$$

and because $(1 - t)$ is fixed in value, whereas as k increases, t^k decreases, (remember that $|t| < 1$), it is an easy exercise to show that

$$\sum_{n=0}^{\infty} t^n = \frac{1}{1-t} \quad \text{as long as} \quad |t| < 1$$

Of course if $|t| > 1$, then $\sum t^n$ is not convergent, we cannot find a sum for the series.

10.6. Convergence of Series

We need to be able to determine whether a given series is convergent or not. However, we have a series that we know is convergent under certain conditions, and we can use that to help us to examine other series. In fact in general, knowing that a particular series is convergent helps us to determine the convergence or otherwise of other series. What we do is to compare the terms of the new series with the terms of the one we already know about.

For instance, suppose we have two series of positive terms, $\sum a_n$ and $\sum b_n$, and that $\sum b_n$ is convergent. Suppose too that after a certain number of terms in each sequence, the term a_n is always greater than zero and less than the term b_n.
(In formal language, there is a certain number N such that for all $n > N, 0 \leq a_n \leq b_n$.)
We can conclude that $\sum a_n$ is convergent. You might stop here and try to prove this assertion for yourself before reading my proof.

Here is the whole thing stated as a theorem and proved:

Theorem 10.6.1. *Suppose we have two series of positive terms, $\sum a_n$ and $\sum b_n$, and that $\sum b_n$ is convergent. Suppose too that there is a certain number N such that for all $n > N, 0 \leq a_n \leq b_n$.*
Then $\sum a_n$ is convergent.

Proof. All we need to look at, is the sequence of sums $A_1 = a_1, A_2 = a_1 + a_2, \ldots$ and so on.

We know that $\sum b_n$ is convergent, so the sequence $B_1, B_2, B_3, \ldots$ has a limit.

Consider A_N and B_N, for some fixed arbitrary large N. Either $A_N \leq B_N$ or $A_N > B_N$.

In the first case, as each subsequent a_n that we will add on to A_N is less than or equal to the corresponding b_n we will be adding to B_N, we can see that for all $n \geq N, A_n \leq B_n$.

So the A_n's are increasing and always bounded above by $\lim B_n$; (since the B_n's are also increasing and converging to this limit). Thus the set of A_n's must have a least upper bound. (Why?) This least upper bound is going to turn out to be $\lim A_n$, but for the moment let us just call it A.

Looking at a diagram, Figure 10.1, makes it obvious.

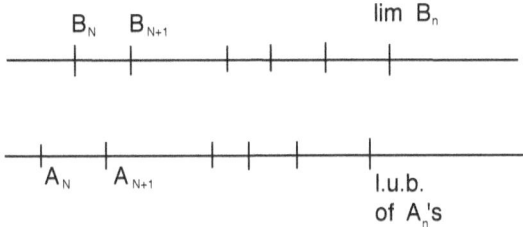

Figure 10.1

Since the A_n's are increasing, by taking sufficiently large values of n, we can get A_n as close to the least upper bound, A, as we like. This means that as A is the least upper bound for the set of A_n's, given any arbitrary real number ε, we can find an n such that $A - A_n < \varepsilon$. In other words, $A_n \in (A - \varepsilon, A + \varepsilon)$.

Hence, given any neighbourhood of A we can find a number N, such that for all $n > N$, A_n lies in that neighbourhood.

So by the definition of a limit, $A = \lim A_n$, and hence $\sum a_n$ is convergent .

If $A_N > B_N$, we can write $A_N = B_N + K$. Then we can use exactly the same argument as above but using $\lim B_n + K$ as the upper bound for the A_n's instead of $\lim B_n$. □

Of course what we have proved only applies to series of positive terms. It would be useful to see what happens if some of the terms in the series are negative and so we look at this case next.

What we can say from the previous theorem, is that if $\sum a_n$ has some negative terms but satisfies the condition that there is a number N such that for all $n \geq N, |a_n| \leq b_n$, then $\sum |a_n|$ is convergent. What we now must prove is that this implies that $\sum a_n$ is convergent. That is not too difficult to prove.

Theorem 10.6.2. *Suppose we have two series, $\sum a_n$ and $\sum b_n$, and that $\sum b_n$ is a convergent series of positive real numbers. Suppose further that for the series $\sum a_n$ some values of $a_n < 0$, and that the series $\sum a_n$ satisfies the condition that there is a number N such that for all $n \geq N, |a_n| \leq b_n$, then $\sum a_n$ is convergent.*

Proof. First of all notice that:

$$\left| \sum_{n=p+1}^{q} a_n \right| \leq \sum_{n=p+1}^{q} |a_n| \qquad (10.1)$$

When a series, $\sum b_n$, converges, then the sequence $B_1, B_2, \ldots$ has a limit. But $\lim B_n$ exists if and only if for any $\varepsilon > 0$ we can find a number N such that for all $p, q > N, |B_p - B_q| < \varepsilon$. This is a simple consequence of the definition of a limit of a sequence and again I will leave the proof to you.

In the case of our series we can now say that $\sum b_n$ converges if and only if for any $\varepsilon > 0$, we can find a number N such that for all $q > p > N, \left| \sum_{n=p+1}^{q} b_n \right| < \varepsilon$

So as $\sum |a_n|$ converges, (by Theorem 10.6.1), then for any $\varepsilon > 0$, there is a number N such that

$$\text{for all } q > p > N, \left| \sum_{n=p+1}^{q} |a_n| \right| < \varepsilon$$

Of course the very outside modulus sign is redundant.
However from equation 10.1 above we have that

$$\left| \sum_{n=p+1}^{q} a_n \right| \leq \sum_{n=p+1}^{q} |a_n|$$

so it follows that $\left|\sum_{n=p+1}^{q} a_n\right| < \varepsilon$ and this, in turn, implies that $\sum a_n$ is convergent.

So if $\sum a_n$ has some negative terms but satisfies the condition that there is a number N such that for all $n \geq N, |a_n| \leq b_n$, then $\sum a_n$ is convergent. □

I have now assembled all the results I need to prove the *ratio test* for the convergence of a series. This is the test that will enable us to see if a given series is convergent.

Theorem 10.6.3. *The Ratio Test.*
Let $\sum a_n$ be a series. If there is a number N such that for all $n \geq N$, the ratio $\left|\frac{a_{n+1}}{a_n}\right| \leq c$ where $c < 1$, then $\sum a_n$ is convergent.

If, on the other hand, there is a number N such that for all $n > N$, the ratio $\left|\frac{a_{n+1}}{a_n}\right| \geq 1$, then $\sum a_n$ is divergent.

Proof. We prove this by first of all, letting $b_n = |a_N| c^{n-N}$ for $n > N$. Since c lies between 0 and 1, $\sum c^k$ converges. (That was our example at the end of section 10.5 earlier). Hence $\sum b_n$ converges also.

Now for $n > N$ we can write

$$|a_n| = \left|\frac{a_n}{a_{n-1}}\right| \cdot \left|\frac{a_{n-1}}{a_{n-2}}\right| \cdot \left|\frac{a_{n-2}}{a_{n-3}}\right| \cdots \left|\frac{a_{N+1}}{a_N}\right| \cdot |a_N|$$

and as each $\left|\frac{a_{r+1}}{a_r}\right|$ is less than or equal to c, and there are n - N of them, the right hand side of the equation is less than or equal to $c^{n-N} \cdot |a_N|$, that is b_n.

So by our result comparing one series with another, we have that $\sum a_n$ is convergent.

To prove the divergence result notice that for all $n \geq N, |a_n| \geq |a_{n-1}|$ so for all $n \geq N, |a_n| \geq |a_N|$.

This implies that the sequence $A_N, A_{N+1}, \ldots A_n$, does not converge, because at each stage you are adding on a number bigger than some fixed positive number, ($|a_N|$), and so the sum will increase without bound. □

In fact we normally use the ratio test in a slightly modified form, known as:

Theorem 10.6.4. *D'Alembert's Ratio Test.*

If $\sum a_n$ is a series, calculate $\lim \left| \frac{a_{n+1}}{a_n} \right|$.

If $\lim \left| \frac{a_{n+1}}{a_n} \right| < 1$, then $\sum a_n$ converges

If $\lim \left| \frac{a_{n+1}}{a_n} \right| > 1$, then $\sum a_n$ diverges

If the limit equals one the series could converge or diverge.

Proof. To prove this, using the ratio test, is easy.

Let $\lim \left| \frac{a_{n+1}}{a_n} \right|$ be K. Then there is a number N such that for all $n \geq N$, $\left| \frac{a_{n+1}}{a_n} \right|$ lies in a neighbourhood of K of chosen radius ε.

So if $K < 1$, let $\varepsilon = \frac{1}{2}(1 - K)$ then for all $n \geq N$, $\left| \frac{a_{n+1}}{a_n} \right| \leq K + \frac{1}{2}(1 - K)$, which is less than 1.

Thus $\sum a_n$ converges.

If $K > 1$, however, given $\varepsilon > 0$, we can find a number N such that for all $n \geq N$, $\left| \frac{a_{n+1}}{a_n} \right|$ lies in a neighbourhood $(K - \varepsilon, K + \varepsilon)$ of K.

Choosing $\varepsilon = \frac{1}{2}(K - 1)$ means that $\left| \frac{a_{n+1}}{a_n} \right| \geq K - \frac{1}{2}(K - 1)$ which is greater than 1.

So $\sum a_n$ diverges.

Do notice that if $K = 1$ we cannot make any statement about $\sum a_n$. □

In this chapter I have assembled a few results about sequences and series leading directly to D'Alembert's Ratio Test. This is because I wish to study in much greater detail a particular kind of series, called a power series, and so I have chosen only those results that will be needed to do this in the next chapter.

In so doing I am aware of the important results that I have ignored. If you are interested in sequences and series then a look at one of the larger standard texts on Analysis will richly reward you, but for our purposes we have done all that we need.

10.7. Exercises

1. (a). For the sequence $1, \frac{1}{2}, \frac{1}{4}, \frac{1}{8}, \frac{1}{16}, \ldots$

   with $A_1 = 1$, $A_2 = 1\frac{1}{2}$, $A_3 = 1\frac{3}{4}$, $A_4 = 1\frac{7}{8}$, $A_5 = 1\frac{15}{16}, \ldots$, prove that

   $$A_n = 2 - \frac{1}{2^{n-1}}$$

   (b). Compute the condition on n for A_n to lie in the open interval $(2-\varepsilon, 2+\varepsilon)$ where ε is an arbitrarily small positive real.

2. Consider the series

   $$\sum_{n=0}^{\infty} t^n = 1 + t + t^2 + t^3 + \ldots$$

   Assume that $|t| < 1$.
   Show that

   $$\sum_{n=0}^{\infty} t^n = \frac{1}{1-t} \quad \text{as long as} \quad |t| < 1$$

3. Suppose we have two series of positive terms, $\sum a_n$ and $\sum b_n$, and that $\sum b_n$ is convergent. Suppose too that there is a certain number N such that for all $n > N, 0 \le a_n \le b_n$.

   Define the two sequences of sums $A_1 = a_1, A_2 = a_1 + a_2, \ldots$ and $B_1 = b_1, B_2 = b_1 + b_2, \ldots$.

   We know that $\sum b_n$ is convergent, so the sequence $B_1, B_2, B_3, \ldots$ has a limit.

   Now consider A_N and B_N, for some fixed arbitrary large N. Assume that $A_N > B_N$, and write $A_N = B_N + K$. Then use the same argument as given in the proof of Theorem 10.6.1 for the case where $A_N \le B_N$, but using $\lim B_n + K$ as the upper bound for the A_n's instead of $\lim B_n$, to prove that the set of A_n's must have a least upper bound and hence that given any neighbourhood of A we can find a number N, such that for all $n > N$, A_n lies in that neighbourhood.

4. Prove that when a series, $\sum b_n$, converges, that is the sequence $B_1, B_2, \ldots$ has a limit, $\lim B_n$ exists if and only if for any $\varepsilon > 0$ we

can find a number N such that for all $p, q > N, |B_p - B_q| < \varepsilon$.

5. Evaluate
$$\sum_{n=2}^{\infty} \frac{1}{n^3 - n}$$

6. Consider the two series:-
$$\sum_{n=1}^{\infty} \frac{1}{n} \text{ and } \sum_{n=1}^{\infty} \frac{1}{(n+1)^2}$$

(a). Show that for both of these series, the D'Alembert Ratio Test is indeterminate.

(b). Show that the first series is divergent and the second is convergent (Hint: To prove the divergence of the first series, consider:

$$\sum_{n=1}^{2^M} \frac{1}{n} = 1 + \frac{1}{2} + \cdots + \frac{1}{2^M}$$

By grouping the fractional terms of this expression into groups of one, two, then four then eight etc, etc. terms and showing that each grouped expression is greater than or equal to 1/2, you can prove that the expression

$$\sum_{n=1}^{2^M} \frac{1}{n}$$

is unbounded as M increases.

Use Theorem 10.6.1 with $b_n = \frac{1}{n(n+1)}$ to prove the convergence of the second series.)

Now for some examples that extend the brief theory given in this chapter.

Sequences

7. Prove that the limit of a sequence is unique, in other words suppose that the sequence $\{a_n\}$ converges to a say, show that the assumption that it also converges to b say, where $a \neq b$ leads to a contradiction.

8. Prove the following Rules for Limits:-
(a). If $\{a_n\}$ and $\{b_n\}$ are sequences such that $\lim_{n\to\infty} a_n = a$ and $\lim_{n\to\infty} b_n = b$, then $\lim_{n\to\infty} (a_n + b_n) = a + b$.
(b). If $\{a_n\}$ and $\{b_n\}$ are sequences such that $\lim_{n\to\infty} a_n = a$ and $\lim_{n\to\infty} b_n = b$, then $\lim_{n\to\infty} (a_n.b_n) = a.b$.
(c). If $\{a_n\}$ is a sequence such that $\lim_{n\to\infty} a_n = a$ and $k \in \mathbb{R}$, then $\lim_{n\to\infty} (k.a_n) = k.a$.
(d). If $\{a_n\}$ is a sequence such that $\lim_{n\to\infty} a_n = a$ where $a \neq 0$, then $\{1/a_n\}$ is a sequence and $\lim_{n\to\infty} (1/a_n) = 1/a$.

A sequence $\{a_n\}$ is called *increasing* if $a_{n+1} \geq a_n$ for all values of n. It is called decreasing if $a_{n+1} \leq a_n$ for all values of n. It is called *monotone* if it is either increasing or else decreasing.

9. (harder) Prove that a monotone sequence which is also bounded (above or below depending on whether the sequence is increasing or decreasing) is convergent.

Series

A series $\sum a_n$ is called *absolutely convergent* if the series $\sum |a_n|$ is convergent.

10. Prove that any absolutely convergent series is convergent.

11. Consider the series:
$$\sum_{n=1}^{\infty} \frac{(-1)^{n+1}}{n} = 1 - \frac{1}{2} + \frac{1}{3} - \frac{1}{4} + \frac{1}{5} - \cdots$$

(a) Show that this series is convergent, even though it is *not* absolutely convergent. (See Question 6 above.)

(b). Let S represent the sum of the series. Now rearrange the terms of the series so that it becomes:-
$$1 - \frac{1}{2} - \frac{1}{4} + \frac{1}{3} - \frac{1}{6} - \frac{1}{8} + \frac{1}{5} - \frac{1}{10} - \frac{1}{12} + \frac{1}{7} - \cdots$$

Consider the sum of the first $3n$ terms, which is:

$$1 - \frac{1}{2} - \frac{1}{4} + \frac{1}{3} - \frac{1}{6} - \frac{1}{8} + \cdots + \frac{1}{2n-1} - \frac{1}{4n-2} - \frac{1}{4n}$$

and rearrange this in the form $A - B - C$ where A, B and C are all sums of positive terms, by placing the first term in A, the second in B, the third in C, the fourth in A again and so on.
Now show that this is equal to:-

$$\frac{1}{2}\left(1 - \frac{1}{2} + \frac{1}{3} - \frac{1}{4} + \frac{1}{5} - \cdots + \frac{1}{2n-1} - \frac{1}{2n}\right)$$

and conclude that the re-arranged series converges to $\frac{1}{2}S$ which is *not* the same as the sum of the original series because $S \neq 0$!!

Hence we have shown that for a series that is convergent but not absolutely convergent it is possible to rearrange the series to give a different sum. Hence in this case, the sum of a series is not unique. However for absolutely convergent series, the sum *is* unique.

Chapter 11
Power Series

11.1. Objectives

In this chapter you meet power series, a powerful way of constructing new functions. You will be able to show that when it converges a power series defines a function that is continuous, differentiable and integrable.
You will be able to determine when a power series does and does not converge.

11.2. New Terms introduced

polynomial function	$f(x) = a_o + a_1 x + a_2 x^2 + \cdots + a_n x^n$
power series	$\sum a_n x^n$
radius of convergence	ρ

11.3. Introduction and Definition

Among the simplest kinds of functions that we have come across are the polynomial functions, like $f(x) = x^7 - 6x^4 + x - 2$ or $g(x) = 1 + 2x - 3x^5$.

In fact any function that we could write in the form $f(x) = a_o + a_1 x + a_2 x^2 + \cdots + a_n x^n$, for some positive whole number n, with the a_i's all real numbers, is a POLYNOMIAL FUNCTION. Other functions like $f(x) = \frac{1}{x}$ or $g(x) = \sin x$ are not polynomial functions. (Yes, I am well aware that I have not actually said yet what the *sine* function is.)

It is fairly obvious that polynomial functions are easier to work with than non-polynomial functions and what we are going to do next is to show how we can define a kind of infinite polynomial function, $a_o + a_1 x + a_2 x^2 + \ldots$ with no last term.

It is this that is called a POWER SERIES, and what I am going to do, is to use the techniques obtained in the last chapter to prove that

145

CHAPTER 11. POWER SERIES

every power series defines a differentiable function, when that power series converges. Some of the proofs in this chapter will be messy and involved but you must make the effort to understand them and to see what is going on.

Definition 11.3.1. *A power series (in x) is a series of the form:*

$$\sum_{n=0}^{\infty} a_n x^n$$

For the sake of convenience, I will normally just use $\sum a_n x^n$ to represent a power series, instead of the fuller notation $\sum_{n=0}^{\infty} a_n x^n$. However I will revert to the fuller notation if there is any possibility of confusion especially over the lowest value of n. (Sometimes n needs to be 1, or even another integer, rather than 0).

11.4. Convergence

So suppose we have a power series, $\sum a_n x^n$, and we want to decide for what values of x it will converge. There are three possibilities.
Firstly, it only converges when $x = 0$. (Trivially every power series will converge when $x = 0$).
Secondly, it will converge for every value of x or:
Thirdly, there is some number ρ such that when $|x|$ is less than ρ the series converges but when $|x|$ is greater than ρ the series diverges.

In this third case, the number ρ is called the RADIUS OF CONVERGENCE of the power series.

How do I know that those are the only three possibilities?

Well consider the set, S of values of x for which $\sum a_n x^n$ converges. 0 belongs to S, obviously. Suppose c belongs to S, for some number c. Then $\sum a_n c^n$ converges.
So the terms $a_n c^n$ must get smaller as n gets bigger (why?), i.e. $\lim a_n c^n = 0$.
So for some number N we have $|a_n c^n| < 1$ for all $n > N$ and this means that

$$|a_n x^n| \text{ which equals } \left|\frac{x^n a_n c^n}{c^n}\right|, \text{ in other words } \left|\frac{x}{c}\right|^n |a_n c^n|$$

by standard algebraic manipulation, is less than

$$\left|\frac{x}{c}\right|^n \text{ for all } n > N$$

11.4. CONVERGENCE

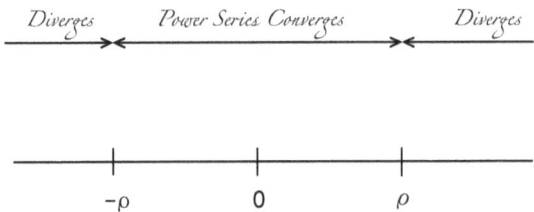

When x = ρ the power series may or may not converge.

Figure 11.1

However we know from section 10.6 that $\sum \left|\frac{x}{c}\right|^n$ converges whenever $\left|\frac{x}{c}\right| < 1$, that is $|x| < |c|$. Thus we have that $\sum |a_n x^n|$ converges for $|x| < |c|$, hence $\sum a_n x^n$ does too.

So we have proved that if the power series for converges for $x = c$ then it also converges for all values of x such that $|x| \leq |c|$.

This means, because we know every power series converges for $x = 0$, that we could have just the sole number 0 in S or we could have all values of x in S or, we have the intermediate condition that some numbers are in S and some are not.
In this third case, by what we have proved above, S must then be bounded so it will have a least upper bound that we will call ρ.

It is not difficult to calculate ρ. Let us see how we can do this. Suppose $\sum a_n x^n$ is a power series for which we want to find out the values of x for which it converges. Then by D'Alembert's Ratio Test we look at

$$\lim \left|\frac{a_{n+1} x^{n+1}}{a_n x^n}\right| \text{ which is equal to } |x| \lim \left|\frac{a_{n+1}}{a_n}\right|$$

Let us write b for $\lim \left|\frac{a_{n+1}}{a_n}\right|$ to simplify our algebra a little bit..

147

The series is convergent if $|x| \cdot b < 1$ that is $|x| < 1/b$, and so will be divergent if $|x| > 1/b$.
This means that ρ must equal $1/b$.

Examples 11.4.1

1. Consider
$$\sum_{n=0}^{\infty} \frac{x^n}{n!}$$
Then b, which is defined as
$$\lim \left|\frac{a_{n+1}}{a_n}\right| = \lim \left|\frac{n!}{(n+1)!}\right| = \lim \left|\frac{1}{n+1}\right| = 0$$

So the series converges for $|x| \cdot 0 < 1$ i.e. for all x.
We sometimes say that the series has an infinite radius of convergence.

2. Look at $\sum_{n=0}^{\infty} n! x^n$. Here
$$b = \lim \left|\frac{a_{n+1}}{a_n}\right| = \lim \left|\frac{(n+1)!}{n!}\right| = \lim |n+1|$$
which just keeps on getting bigger!
So $1/b$ will be 0, thus the series only converges when x = 0.

3. As an example of the in between case look at
$$\sum_{n=0}^{\infty} \frac{x^n}{(n+1)(n+2)}$$

Then $b = lim \left|\frac{a_{n+1}}{a_n}\right| = lim \left|\frac{(n+2)(n+3)}{(n+1)(n+2)}\right| = lim \left|\frac{n+3}{n+1}\right|$

To evaluate this limit we use a little low cunning and divide top and bottom by n, so
$$b = \lim \left|\frac{1+\frac{3}{n}}{1+\frac{1}{n}}\right| = 1$$
So this third power series has radius of convergence 1.

11.5. Continuity of Power Series

This is the first of our two main results, namely that any power series defines a continuous function where that power series converges, or more formally:

Theorem 11.5.1. *Let $\sum a_n x^n$ be a power series with radius of convergence ρ. Then the function $f(x)$ defined by letting $f(x) = \sum a_n x^n$ is defined and continuous on the open interval $(-\rho, \rho)$.*

Proof. We will prove this by choosing $0 < c < \rho$ so $\sum a_n x^n$ converges for $x = c$.
Then I prove that $f : [0, c] \longrightarrow \mathbb{R}$ defined as above is continuous.

The proof of continuity for f from $[-c, 0]$ to $\mathbb{R}$ follows by replacing $\sum a_n x^n$ by $\sum (-1)^n a_n x^n$.

Suppose then $\sum a_n x^n$ converges for $x = c > 0$, and that we have a point x_o belonging to $[0, c]$. Given an $\varepsilon > 0$, to prove that f is continuous at x_o we must prove the existence of a neighbourhood N_{x_0} such that when x is in N_{x_0}, $f(x)$ lies in $(f(x_o) - \varepsilon, f(x_o) + \varepsilon)$.

<u>Step 1</u>. We find a number k_o such that for all $m > k > k_o$

$$\left| \sum_{k+1}^{m} a_n c^n \right| < \frac{\varepsilon}{3}$$

As $\sum a_n x^n$ converges we can do this, as explained in the last chapter.

<u>Step 2</u>. We now show that for all x in $[0,c]$ and $m > k > k_o$

$$\left| \sum_{k+1}^{m} a_n x^n \right| < \frac{\varepsilon}{3}$$

Consider x in $[0, c]$. Let $b_n = a_n c^n$ and $y = x/c$.
Then we can write $\sum a_n x^n$ as $\sum b_n y^n$ where y is in $[0,1]$.
Now we use some fiendish ingenuity and say that

$$\left|\sum_{k+1}^{m} b_n y^n\right| =$$

$$|b_{k+1}(y^{k+1} - y^{k+2}) + (b_{k+1} + b_{k+2})(y^{k+2} - y^{k+3}) +$$
$$(b_{k+1} + b_{k+2} + b_{k+3})(y^{k+3} - y^{k+4}) + \ldots$$
$$+ (b_{k+1} + \ldots b_{m-1})(y^{m-1} - y^m)$$
$$+ (b_{k+1} + \ldots b_m)y^m|$$

From step 1, we have $\left|\sum_{k+1}^{m} b_n\right| < \frac{\varepsilon}{3}$, so we can write

$$\left|\sum_{k+1}^{m} b_n y^n\right| <$$
$$\frac{\varepsilon}{3}(y^{k+1} - y^{k+2}) + \frac{\varepsilon}{3}(y^{k+2} - y^{k+3}) + \ldots$$
$$+ \frac{\varepsilon}{3}(y^{m-1} - y^m) + \frac{\varepsilon}{3}y^m$$

because each of the sums of b_i's is less than or equal to $\sum_{k+1}^{m} b_n$.

However, this new expression simplifies down to:

$$\frac{\varepsilon}{3}y^{k+1}$$

and as y lies between 0 and 1, this is less than $\varepsilon/3$.

So $\left|\sum_{k+1}^{m} a_n x^n\right| < \frac{\varepsilon}{3}$, for all x in $[0,c]$ and $m > k > k_o$.

<u>Step 3</u>. As $f(x) - \sum_0^n a_n x^n$ can be written as $lim\left(\sum_{k+1}^{m} a_n x^n\right)$, the limit being taken as m gets larger and larger, in other words m is tending to infinity, we can conclude that

$$\left|f(x) - \sum_0^k a_n x^n\right| < \frac{\varepsilon}{3}$$

for all x in $[0,c]$ and $k > k_o$.

<u>Step 4</u>. Consider the polynomial function $S_k(x) = \sum_0^k a_n x^n$. Since $S_k(x)$ is a polynomial function, it is continuous, hence there is

some number δ such that when $|x - x_0| < \delta$ then $|(S_k(x) - S_k(x_0)| < \varepsilon/3$.
(The reason for all these $\varepsilon/3$'s will become apparent in a moment.)

Now we look at $|f(x) - f(x_0)|$.
We can write $|f(x) - f(x_0)|$ as $|(f(x) - S_k(x)) + (S_k(x) - S_k(x_0)) + (S_k(x_0) - f(x_0))|$ which is certainly less than or equal to $|f(x) - S_k(x)| + |(S_k(x) - S_k(x_0))| + |(S_k(x_0) - f(x_0)|$

However $|f(x) - S_k(x))| < \varepsilon/3$ by Step 3, $|(S_k(x) - S_k(x_0))| < \varepsilon/3$ by the statement at the beginning of step 4 and $|(S_k(x_0) - f(x_0)| < \varepsilon/3$ by step 3 again.

So $|f(x) - f(x_0)| < \varepsilon$ and that means that $f(x)$ lies in $(f(x_0) - \varepsilon, f(x_0) + \varepsilon)$, thus f is continuous on $[0,c]$. $\square$

So, for example,

$$\sum_{n=0}^{\infty} \frac{x^n}{(n+1)(n+2)}$$

not only converges whenever x is in (-1,1) but does so in such a way as to define a continuous function. Of course, actually calculating the values it assumes and so plotting a graph of the function would involve a great deal of work.

In the last chapter we shall examine three special power series in great detail because they define very special functions, but before we do that I need to prove that a power series defines not just a continuous function, but also a differentiable one.

11.6. Differentiability of Power Series

Before I try to prove that a power series defines a differentiable function, I need a result that relates the convergence of one power series to certain other power series that are similar to the first.

Theorem 11.6.1. *Suppose that $\sum_{n=0}^{\infty} a_n x^n$ has a radius of convergence ρ then so will the following three power series:*

$$\sum_{n=0}^{\infty} (n+1) a_{n+1} x^n$$

CHAPTER 11. POWER SERIES

$$\sum_{n=1}^{\infty} \frac{a_{n-1}}{n} x^n$$

$$\sum_{n=0}^{\infty} (n+2)(n+1) a_{n+2} x^n$$

Proof. I will prove this for the first of the three series and leave the other two to you.
(The third one follows from the first fairly easily, just try writing b_n for $(n+1)a_{n+1}$ and apply the first result again, the second one needs to be worked out in detail.)
Let ρ' be the radius of convergence of $\sum (n+1)a_{n+1}x^n$ and suppose that $|x| < \rho'$ so
$\sum |(n+1)a_{n+1}x^n|$ converges. As $n+1$ is positive we can say that $\sum (n+1)|a_{n+1}x^n|$ converges and hence so does $\sum |a_{n+1}x^n|$, as each term in this last series is smaller than the corresponding term in the previous one. If we multiply the series by $|x|$ this will not affect its convergence (why not?), so $|x| \sum |a_{n+1}x^n|$ converges.
We can rewrite this and deduce that $\sum |a_{n+1}x^{n+1}|$ converges and thus $|x|$ must be less than ρ.
So I have shown that $\rho' \leq \rho$.

I will now prove the converse. Take $|x|$ less than some positive number K which is in turn less than ρ. This means that both $\sum a_n x^n$ and $\sum a_n K^n$ converge, and so $\sum |a_n x^n|$ converges. As the terms in a convergent series can be made as small as we like if we go far enough along the series, in other words choose a big enough n, we can say that for all n greater than some n_0, $|a_n K^n|$ is less than εK where ε is a chosen small number.

Now $\sum |(n+1)a_{n+1}x^n|$ is equal to $\sum \left| a_{n+1} K^n (n+1) \left(\frac{x}{K} \right)^n \right|$

and since, once $n > n_0$, $|a_{n+1} K^n| < \varepsilon$, this series is less than

$$\sum (n+1) \left(\frac{x}{K} \right)^n,$$

which can be easily shown to converge by D'Alembert's Ratio Test. Thus $\sum (n+1)a_{n+1}x^n$ converges so $|x| < \rho'$ and $\rho \leq \rho'$.

11.6. DIFFERENTIABILITY OF POWER SERIES

We have now proved that $\rho' = \rho$.

$\square$

Let us briefly look at an example.

Consider the power series $\sum_{n=0}^{\infty} x^n$. We have already studied this in the previous chapter and we know that this series has radius of convergence 1.

We also know that where it converges it agrees with the continuous (and differentiable) function $f(x) = \frac{1}{1-x}$. So we could write:

$$\frac{1}{1-x} = \sum_{n=0}^{\infty} x^n = 1 + x + x^2 + \ldots\ldots\ldots \text{ for } |x| < 1.$$

The result I have just proved tells us that the series $\sum_{n=0}^{\infty} (n+1)x^n$ converges for $|x| < 1$ as well.

This new series looks like the result of differentiating that first series term by term.

Is it therefore equal to $f'(x)$, namely $\frac{1}{1-x^2}$?

The answer is yes, but it still has to be proved!

We can also conclude that the series $\sum_{n=1}^{\infty} \frac{1}{n} x^n$ converges when $|x| < 1$ (notice we must prevent n=0).

This series looks like the anti-derivative of the original series, and again it is, but once again this needs to be proved.

Now I am going to prove that if $\sum a_n x^n$ has radius of convergence ρ, then it defines a differentiable function $f(x)$ such that $f'(x)$ which will equal $\sum (n+1)a_{n+1} x^n$ also has radius of convergence ρ. We had better state this formally as a theorem.

Theorem 11.6.2. *Suppose that $\sum a_n x^n$ has radius of convergence ρ and let x_0 be a point chosen so that $|x_0| < \rho$. Then the function $f(x)$ defined by letting $f(x) = \sum a_n x^n$ is differentiable at x_0 and $f'(x_0) = \sum (n+1)a_{n+1} x_0^n$.*

Note that I already know that this last expression will be convergent by Theorem 11.6.1

Proof. Choose an x such that $|x| < \rho$ as well, then $f(x) - f(x_0) = \sum a_n(x^n - x_0^n)$ which can be factorised as:

$$(x - x_0) \sum a_n(x^{n-1} + x^{n-2}x_0 + \cdots + x_0^{n-1})$$

Define the chord slope function $p_{x_0}(x)$ by $\sum a_n(x^{n-1} + x^{n-2}x_0 + \cdots + x_0^{n-1})$.

Obviously $p_{x_0}(x_0) = \sum na_n x_0^{n-1}$ so this will be $f'(x_0)$. (Compare this with the form given in the theorem.)

All we now have to prove is that $p_{x_0}(x)$ is continuous at x_0.
To show this I take a K such that both $|x| < K < \rho$ and $|x_0| < K < \rho$, then consider $|p_{x_0}(x) - p_{x_0}(x_0)|$

If I can show that this is less than some constant multiple of $|x - x_0|$ then I will have proved what I want.

First of all:

$$|p_{x_0}(x) - p_{x_0}(x_0)| = \left|\sum a_n \left[(x^{n-1} - x_0^{n-1}) + (x^{n-2}x_0 - x_0^{n-1}) + \cdots \right.\right.$$
$$\left.\left. \cdots + (xx_0^{n-2} - x_0^{n-1}) + (x_0^{n-1} - x_0^{n-1})\right]\right|$$

which is less than or equal to

$$\sum |a_n| \left[\left|x^{n-1} - x_0^{n-1}\right| + |x_0|\left|x^{n-2} - x_0^{n-2}\right| + \cdots \right.$$
$$\left. \cdots + \left|x_0^{n-2}\right| |x - x_0| + 0\right]$$

Now the Mean Value Theorem tells us that we can find a y between x and x_0, such that $|x^n - x_0^n| = |ny^{n-1}| |x - x_0|$.
Since y lies between x and x_0, $|y| < K$ thus $|x^n - x_0^n| \leq |nK^{n-1}| |x - x_0|$
Using this result, and the fact that $x_0 < K$, we can see that the last expression will be less than or equal to

$$\sum |a_n| \left[(n-1)K^{n-2} + K(n-2)K^{n-3} + \ldots + K^{n-2}\right] |x - x_0|$$

which is just

$$\sum |a_n|K^{n-2}[(n-1)+(n-2)+\ldots+2+1]|x-x_0|$$

that is

$$\sum |a_n|K^{n-2}\frac{n(n-1)}{2}|x-x_0|$$

using the formula for the sum of the first n integers.

This series can be written as $|x-x_0|\frac{1}{2}\sum n(n-1)a_n K^{n-2}$ (Why?)
that is $|x-x_0|M$ where $2M$ is defined by the series $\sum n(n-1)a_n K^{n-2}$ which is convergent by Theorem 11.6.1

So p_{x_0} is continuous at x_0, hence f is differentiable there. Since x_0 was any point chosen within the radius of convergence ρ, f is differentiable everywhere within the radius of convergence of f.

□

In terms of our example earlier it means that $\sum_{n=0}^{\infty}(n+1)x^n$ converges on (-1, 1) to $f'(x)$, that is

$$\frac{1}{1-x^2}$$

11.7. Integration of Power Series

We can use this result to show that we can also *integrate* a power series term by term, within its radius of convergence.

We know from Theorem 11.6.1 that if $\sum a_n x^n$ converges for $|x| < \rho$ then so does

$$\sum_{n=1}^{\infty}\frac{a_{n-1}}{n}x^n$$

Thus this second series defines a differentiable function, let's call it $F(x)$, and by Theorem 11.6.2 we know that $F'(x) = \sum a_n x^n$.

This means that if we choose any two points s and t that belong to $(-\rho,\rho)$ then both $\sum_{n=1}^{\infty} \frac{a_{n-1}}{n} t^n$ and $\sum_{n=1}^{\infty} \frac{a_{n-1}}{n} s^n$ will converge, so both $F(t)$ and $F(s)$ are properly defined.

Then as we know that $F(t) - F(s)$ exists, from the First Fundamental Theorem of Calculus we have that $F(t) - F(s)$ must be the integral of F' over the interval $[s,t]$ that is the integral of $\sum a_n x^n$ over that interval.

This last paragraph is somewhat convoluted and I advise you to spend a few minutes convincing yourself that it is true.

We have now assembled all the basic tools and techniques of Analysis and in the next chapter I will attempt to use them to define and investigate those functions called the elementary functions, which are the sine, cosine and exponential functions. I will also try to relate these functions to their simpler geometric definitions.

11.8. Exercises

1. Suppose that $\sum_{n=0}^{\infty}$ has a radius of convergence ρ prove that the following two power series will also converge within the same radius:

(a)
$$\sum_{n=1}^{\infty} \frac{a_{n-1}}{n} x^n$$

(b)
$$\sum_{n=0}^{\infty} (n+2)(n+1)a_{n+2} x^n$$

2. Prove that if we multiply a convergent power series by $|x|$ this will not affect its convergence.

3. Find the radius of convergence of the following power series:-

(a)
$$\sum_{n=1}^{\infty} \left(\frac{x}{n}\right)^n$$

(b)
$$\sum_{n=0}^{\infty} \frac{x^n}{n!}$$

(c)
$$\sum_{n=0}^{\infty} (-1)^n \frac{x^{2n}}{(2n)!}$$

(d)
$$\sum_{n=0}^{\infty} (-1)^n \frac{x^{2n+1}}{(2n+1)!}$$

(e)
$$\sum_{n=1}^{\infty} \frac{(n!)^2}{(2n)!} x^n$$

(f)
$$\sum_{n=1}^{\infty} \frac{x^n}{\sqrt{n}}$$

4. Prove that the power series $\sum_{n=0}^{\infty} n^2 x^n$ converges for $|x| < 1$ and show that its sum is equal to

$$\frac{x(1+x)}{(1-x)^3}$$

Chapter 12
The Elementary Functions

12.1. Objectives

By the end of this chapter you will understand the formal definitions of the exponential and basic trigonometric functions. You will have proved their basic properties and seen how they relate to the naive definitions you have met before. You will also have met methods for calculating π to any required degree of accuracy.

In the exercises you will extend this theory to defining the natural logarithm and principal inverses of the trigonometric functions.

12.2. New Terms introduced

exponential function	exp
sine function	sin
cosine function	cos
tangent function	tan
inverse to tangent function	arctan
natural logarithm function	ln

12.3. Definitions

We have spent a great deal of time in the last couple of chapters on definitions and proofs with very little in the way of examples. This last chapter is intended to correct this by being totally made up of examples. We will look at three functions, all defined by power series, and, using the results that we have proved, we will deduce their properties. The three functions are already familiar from elementary trigonometry and calculus namely the sine, the cosine and the exponential functions.

So let's start by defining these functions in terms of power series.

CHAPTER 12. THE ELEMENTARY FUNCTIONS

Definition 12.3.1. *For all real numbers x, define:*

$$\exp x = e^x = \sum_{n=0}^{\infty} \frac{x^n}{n!} = 1 + x + \frac{x^2}{2!} + \ldots$$

$$\cos x = \sum_{n=0}^{\infty} (-1)^n \frac{x^{2n}}{(2n)!} = 1 - \frac{x^2}{2!} + \frac{x^4}{4!} - \ldots$$

$$\sin x = \sum_{n=0}^{\infty} (-1)^n \frac{x^{2n+1}}{(2n+1)!} = x - \frac{x^3}{3!} + \frac{x^5}{5!} - \ldots$$

Each of these series converges for all values of x, this is easy to see using D'Alembert's ratio test and was an exercise from the last chapter.

(Also, please note that I will use the notations $\exp x$ and e^x interchangeably depending on which is most convenient. The number e^1 is denoted by e, after the Swiss mathematician Euler.)

Now let's see if we can derive some easy results that agree with our previous understanding of these three functions.

First of all, straight from the definitions we can see that:

$$\cos(-x) = \cos x, \ \sin(-x) = -\sin x, \ e^0 = 1, \ \cos 0 = 1, \ \sin 0 = 0.$$

Secondly, with a little bit of care, it should be easy for you to prove that

$$\lim_{x \to 0} \frac{\sin x}{x} = 1$$

and so I leave this for you as an exercise.

We can define new trigonometric functions in the standard way, for instance, we will define

$$\tan x \text{ as } \frac{\sin x}{\cos x}, \text{ as long as } \cos x \neq 0$$

In the last chapter we learnt that we can differentiate a power series term by term, and using Theorem 11.6.2 now, we have:

Theorem 12.3.1. *For all values of x,*

$$\exp' x = \exp x, \cos' x = -\sin x \text{ and } \sin' x = \cos x$$

12.4. SOME PROPERTIES OF THE EXPONENTIAL FUNCTION

I will only actually go through all the stages of the proof for cos′, as this is the most complicated one of the three, the other two are proved in exactly the same way and I leave those as exercises for you.

Proof. From Theorem 11.6.2 in the last chapter and the definition above that

$$\cos x = \sum_{n=0}^{\infty} (-1)^n \frac{x^{2n}}{(2n)!}$$

we can say that:

$$\cos x = 1 + \sum_{n=1}^{\infty} (-1)^n \frac{x^{2n}}{(2n)!}$$

so $\cos' x = 0 + \sum_{n=1}^{\infty} (-1)^n (2n) \frac{x^{2n-1}}{(2n)!} = \sum_{n=1}^{\infty} (-1)^n \frac{x^{2n-1}}{(2n-1)!}$

Letting $m = n - 1$, so $2n - 1 = 2m + 1$ we can write

$$\cos' x = \sum_{m=0}^{\infty} (-1)^{m+1} \frac{x^{2m+1}}{(2m+1)!} = -\sum_{m=0}^{\infty} (-1)^m \frac{x^{2m+1}}{(2m+1)!} = -\sin x$$

□

Notice that if I had not written $\cos x$ as

$$1 + \sum_{n=1}^{\infty} (-1)^n \frac{x^{2n}}{(2n)!}$$

at the beginning of the proof, I would have run into some difficulty on differentiating. That step was really just applied low cunning!

12.4. Some Properties of the Exponential Function

One of the main properties of the exponential function is its additive property which I will prove next.

Theorem 12.4.1. *For all real numbers c and d, $e^{c+d} = e^c . e^d$.*

Proof. To prove this we choose any real number a and let $f(x) = e^x.e^{a-x}$.

Differentiating, using the product rule, we get
$f'(x) = e^x.e^{a-x} + e^x.(-e^{a-x}) = 0$.

So $f(x)$ must be constant on the real line. But $f(0) = e^a$ hence we have shown that $e^a = e^x.e^{a-x}$.

Replacing x by c and a-x by d gives $e^{c+d} = e^c.e^d$ as we require. □

It follows immediately that e^x is never zero for any value of x and that e^{-x} is equal to $1/x$.
(Why does it follow? Another exercise for you!)

From the series for e^x we see that the exponential function is both strictly positive and strictly increasing for $x > 0$. Again, why?

Also from the series, $x^{-n}e^x > \frac{x}{(n+1)!}$ (Here I am just saying that e^x is larger than any one of the terms in its power series.)

As $x \to \infty$, then $\frac{x}{(n+1)!} \to \infty$, as well, so we deduce that as $x \to \infty$, then $x^{-n}e^x \to \infty$

Now as $e^{-x} = 1/e^x$ you can deduce that $x^n e^{-x} \to 0$ as $x \to \infty$ and also that for $x < 0$, e^x is both strictly positive and strictly increasing, hence it is both of those on the *whole* of the real numbers.

12.5. Some Properties of the Trigonometric Functions

We now turn our attention to the trigonometric functions. Firstly I will prove the addition formulae.

Theorem 12.5.1. *For every a and b in the reals*

$$\cos(a+b) = \cos a \cos b - \sin a \sin b$$

and

$$\sin(a+b) = \sin a \cos b + \cos a \sin b$$

Proof. The proof of this is very similar to the proof of the exponential addition formulae, so I will only prove the result for cos, leaving you to complete the proof for sin.

Choose c in $\mathbb{R}$ and let

$$f(x) = \cos x \cos(c-x) - \sin x \sin(c-x)$$

for x in $\mathbb{R}$. Differentiating we have:

$$\begin{aligned}f'(x) = &-\sin x \cos(c-x) + \cos x \sin(c-x)\\&-\cos x \sin(c-x) + \sin x \cos(c-x)\\= &\,0\end{aligned}$$

for all real x so f is constant on the reals.
Hence $f(x) = f(0) = \cos c$ for all values of x.
Letting $x = a, c - x = b$ gives the final result. □

As a consequence of this theorem, if we put $b = -a$, and use the results that $\cos(-x) = \cos x, \sin(-x) = -\sin x$ and $\cos 0 = 1$ we get the result that $\cos^2 a + \sin^2 a = 1$ for any real number a.
Hence both $\sin x$ and $\cos x$ lie in the interval [-1, 1] whatever the value of x.

Of course both you and I know exactly what the sine and cosine are in geometric terms and all the results that I have proved are no surprise at all. However in order to finally tie up the power series definition with our previous knowledge we are going to need to involve that number π somewhere.
The next section does just that.

12.6. Pi, $\pi, 3.14159\ldots$

Let's begin by establishing the existence of π.

Theorem 12.6.1. *There is a positive number, which we will call π, such that $\cos(\pi/2) = 0, \sin(\pi/2) = 1$ and $\cos$ is strictly decreasing, and $\sin$ is strictly increasing on the interval $[0, \pi/2]$.*

Proof. From the Mean Value Theorem, Theorem 6.4.2, we can say that if f is a differentiable function such that $f'(t) \geq 0$ for $0 \leq t \leq x$, then $f(x) \geq f(0)$. We use this result repeatedly in what follows.

Now as $\cos x$ lies in [-1, 1], it follows that for all $x \geq 0, 1 - \cos x \geq 0$. But $1 - \cos x$ is the derivative of $x - \sin x$, so $x - \sin x \geq 0$ for $x \geq 0$, by the above result.

However $x - \sin x$ is in its turn, the derivative of $\frac{1}{2}x^2 + \cos x$, so $\frac{1}{2}x^2 + \cos x \geq \cos 0 = 1$ for $x \geq 0$, that is $\frac{1}{2}x^2 - 1 + \cos x \geq 0$, for $x \geq 0$.

Lets repeat this process twice more then we get first that

$$\sin x \geq x - \frac{1}{6}x^3 \text{ for } x \geq 0$$

and then

$$\cos x \leq 1 - \frac{1}{2}x^2 + \frac{1}{24}x^4 \text{ for } x \geq 0$$

Let us take $x = \sqrt{3}$. then we will get that

$$\cos \sqrt{3} \leq 1 - \frac{3}{2} + \frac{9}{24} = -\frac{1}{8}$$

However, cos is a continuous function, (because it is defined by a power series that converges everywhere) so the Intermediate Value Theorem, Theorem ??, tells us that there is some number between 0 and $\sqrt{3}$ with its cosine equal to zero.

I will call this number $\frac{\pi}{2}$.

$$\text{So } 0 < \frac{\pi}{2} < \sqrt{3} \text{ and } \cos(\pi/2) = 0$$

Notice that when $0 < x \leq \sqrt{3}$ we have

$$\sin x \geq x\left(1 - \frac{1}{6}x^2\right) \geq \frac{1}{2}x > 0$$

Now we use the Mean Value Theorem to tell us that cos must be strictly decreasing on $\left[0, \frac{\pi}{2}\right]$ and so strictly positive on $\left[0, \frac{\pi}{2}\right)$.

Make sure that you follow this argument.

Using the Mean Value Theorem again, we can say that sin is strictly increasing on $\left[0, \frac{\pi}{2}\right]$. I also know that for any real number a, $\cos^2 a + \sin^2 a = 1$, so, in particular, $\sin^2(\pi/2) = 1$. As $\sin(\pi/2) > 0, \sin(\pi/2)$ must equal 1.
I have now proved the theorem. □

Using the addition formulae for sin and cos we (actually you) can calculate $\sin \pi$, $\cos \pi$, $\sin 2\pi$ and $\cos 2\pi$. What are they?

We can also see that $\cos(x+2\pi) = \cos x$, $\sin(x+2\pi) = \sin x$, $\cos\left(\frac{\pi}{2} - x\right) = \sin x$ and $\sin\left(\frac{\pi}{2} - x\right) = \cos x$, for all values of x.

Hence $\sin x = 0$ if and only if $x = n\pi$ and $\cos x = 0$ if and only if $x = (n + \frac{1}{2})\pi$, where n is any integer.

To show this, notice that $\sin x \neq 0$ for $0 < x < \pi/2$, and as $\sin(\pi/2 + x) = \cos x$, it follows that $\sin x \neq 0$ for $0 < x < \pi$.
Also we know that $\sin(-x) = -\sin x$ so the only values of x, for which $\sin x$ could be zero, in the interval $[-\pi,\pi]$, are $-\pi$, 0, and π. As $\sin(x+ 2\pi) = \sin x$ we can deduce that $\sin x = 0$ if and only if $x = n\pi$. To prove the second result all I need is that $\cos x = \sin(x + \pi/2)$ and if follows immediately.

I have now deduced all the main properties of the trigonometric functions, but it seems a little unsatisfactory to leave π just as some number between 0 and $2\sqrt{3}$. So I will finish this section by showing how to evaluate π, by finding a series that converges to it.

Remember I defined

$$\tan x \text{ as } \frac{\sin x}{\cos x}, \cos x \neq 0$$

On the open interval $\left(-\frac{\pi}{2}, \frac{\pi}{2}\right)$, tan will be strictly increasing, because

$$\tan' x = \frac{1}{\cos^2 x}$$

which is greater than 0 for x in $\left(-\frac{\pi}{2}, \frac{\pi}{2}\right)$. Furthermore tan takes all real values for x in that range. (Prove it!)
Thus if we consider tan as a function from just $\left(-\frac{\pi}{2}, \frac{\pi}{2}\right)$ to $\mathbb{R}$ then there exists an inverse function for this restricted tangent function, and we call this inverse function "arctan". This is known as "the principal inverse for tan", because a true inverse for the tangent function as a whole does not exist.

So
$$\arctan : \mathbb{R} \longrightarrow \left(-\frac{\pi}{2}, \frac{\pi}{2}\right)$$

Now
$$\tan' x = \frac{1}{\cos^2 x} = 1 + \tan^2 x \text{ for } x \text{ in } \left(-\frac{\pi}{2}, \frac{\pi}{2}\right)$$
so by the rule for differentiating inverse functions, Theorem 5.3.1

$$\arctan' x = \frac{1}{1+x^2}, \text{ for all } x \text{ in } \mathbb{R}$$

So by the fundamental theorem of calculus we can write

$$\arctan x = \int_0^x \frac{1}{1+t^2}\, dt, \text{ for } x \text{ in } \mathbb{R}$$

Now we can calculate a power series expansion for arctan x, as long as x lies in [-1, 1], so the power series will have radius of convergence 1.

Theorem 12.6.2. *If x lies in [-1, 1] then*

$$\arctan x = \sum_{n=0}^{\infty} (-1)^n \frac{x^{2n+1}}{2n+1}$$

Proof. Consider $s(t) = 1 - t^2 + t^4 - \ldots + (-1)^n t^{2n}$
So
$$\left(1+t^2\right).s\left(t\right) = 1 + (-1)^n t^{2n+2}$$

and, dividing both sides of that equation by $1 + t^2$, and then integrating we will get:

$$\arctan x = \int_0^x \frac{1}{1+t^2}\, dt = \int_0^x s(t)\, dt + r(x)$$

where $r(x)$ is defined as

$$(-1)^{n+1} \int_0^x \frac{t^{2n+2}}{1+t^2}\, dt$$

Now if $0 \le t \le x$, then $\dfrac{t^{2n+2}}{1+t^2} \le t^{2n+2}$

So $|r(x)| \le \left| \int_0^x t^{2n+2}\, dt \right| = \dfrac{|x|^{2n+3}}{2n+3}$

Now if $|x| \le 1$, letting $n \to \infty$ makes $r(x) \to 0$. Hence as $n \to \infty$, with $|x| \le 1$

$$\arctan x = \int_0^x \sum_{n=0}^{\infty} (-1)^n t^{2n}\, dt = \sum_{n=0}^{\infty} (-1)^n \frac{x^{2n+1}}{2n+1}$$

$\square$

To finish our evaluation of π note from the addition formula that $\cos 2x = \cos^2 x - \sin^2 x$ so

$$\cos\frac{\pi}{4} = \sin\frac{\pi}{4} \text{ which means that } \tan\frac{\pi}{4} = 1$$

Thus

$$\frac{\pi}{4} = \arctan 1 = \sum_{n=0}^{\infty}(-1)^n \frac{1}{2n+1} = 1 - \frac{1}{3} + \frac{1}{5} - \ldots$$

and so we can compute π as accurately as we want.

In practice the series I have found converges too slowly to be of real use in computing π very accurately, and to perform this calculation other, similar, series are used. Each one is derived from the series for arctan however.

Two popular examples are

$$\frac{\pi}{4} = \arctan\frac{1}{2} + \arctan\frac{1}{3} = \sum_{n=0}^{\infty} \frac{(-1)^n}{2n+1}\left(\frac{1}{2^{2n+1}} + \frac{1}{3^{2n+1}}\right)$$

or

$$\frac{\pi}{4} = 4\arctan\frac{1}{5} - \arctan\frac{1}{239}$$

I have shown enough of the properties of sin and cos to demonstrate that they are the familiar functions of trigonometry and calculus, and also to illustrate some of the results I have proved in the previous chapters. "Do you mean all the work in this book was just to define sine and cosine? What was wrong with opposite over hypoteneuse etc.?" Well the naive definition of sin and cos was no definition at all, as it depended on vague concepts such as "angles". Also in the elementary treatment of sin and cos all the difficult points are neatly skated over whilst I have presented a fully rigorous treatment with no loopholes!

There are however, many things I have not looked at, such as the principal inverses to sin and cos, namely arcsin and arccos, the other trigonometric functions and so on, but there is a limit to the amount that I can write and you can read and all these other things can be found in other, larger, text books. This is only meant to be beginning steps in analysis not all the analysis there ever was!

CHAPTER 12. THE ELEMENTARY FUNCTIONS

One thing that I have omitted that perhaps I should not, is the inverse function to exp. So I will leave that as an exercise.

12.7. Exercises

1. Prove that
$$\lim_{x \to 0} \frac{\sin x}{x} = 1$$

2. Prove that for all values of x,
$$\exp' x = \exp x \text{ and } \sin' x = \cos x$$

3. Show that e^x is never zero for any value of x and that e^{-x} is equal to $1/e^x$.

4. From the series for e^x prove that the exponential function is both strictly positive and strictly increasing for $x > 0$.

5. For every a and b in the reals, prove that
$$\sin(a+b) = \sin a \cos b + \cos a \sin b$$

6. Using the addition formulae for sin and cos calculate $\sin \pi$, $\cos \pi$, $\sin 2\pi$ and $\cos 2\pi$.

7. Prove that for x in the range $\left(-\frac{\pi}{2}, \frac{\pi}{2}\right)$, the tangent function takes all real values.

8. Prove from the addition formula, Theorem 12.5.1, that $\cos 2x = \cos^2 x - \sin^2 x$

9. Show that exp has an inverse function defined on $(0, \infty)$. Call this function ln (this is a shorthand for *natural logarithm*). Using the rule for differentiating inverse functions find $\ln' x$. Hence find where ln increases or decreases, and is positive, zero and negative.

10. Show that the power series

$$\sum_{n=1}^{\infty} (-1)^{n-1} \frac{x^n}{n}$$

converges to $\ln(1+x)$, provided that $-1 < x \leq 1$. (Hint: Consider $\sum_{n=0}^{\infty} (-1)^n x^n$, find its radius of convergence and to what it converges on that radius.)

11.

(a). Prove that on the interval $\left[-\frac{\pi}{2}, \frac{\pi}{2}\right]$, the sine function is strictly increasing and continuous and show that the image of this interval under the sine function is $[-1, 1]$

Hence we can define an inverse for the sine function restricted to the interval $\left[-\frac{\pi}{2}, \frac{\pi}{2}\right]$, called the *arcsine* function, so:

$$\arcsin : [-1, 1] \longrightarrow \left[-\frac{\pi}{2}, \frac{\pi}{2}\right]$$

(b). Calculate $\arcsin' x$

(c) By considering the power series

$$\sum_{n=0}^{\infty} \frac{(2n)!}{4^n (n!)^2} x^{2n}$$

and determining both its radius of convergence and to what it converges, find a power series that converges to $\arcsin x$ everywhere that the arcsine function is defined.

Bibliography

[Bin81] K.G. Binmore, *The foundations of analysis: A straightforward introduction, book 2 topological ideas*, Cambridge University Press, 1981.

[Fow73] D.H. Fowler, *Introducing real analysis*, Transworld Publishers Ltd., 1973.

[MR68] R.M.F. Moss and G.T. Roberts, *A preliminary course in analysis*, Chapman and Hall Ltd., 1968.

[Sch69] R.L.E. Schwarzenberger, *Elementary differential equations*, Chapman and Hall, 1969.

[Spi06] M. Spivak, *Calculus third edition*, Cambridge University Press, 2006.

BIBLIOGRAPHY

Index

π, 163

arctan, 166
average
    uniqueness, 102
Average value of a function, 108
Average value of the derivative of a function over a closed interval, 99

boundary points of an interval, 20
bounded function
    bounded above
        bounded below, 54
bounded interval, 19

cartesian product of two sets, 10
Chain Rule, 78, 86
chord slope function, 66
closed interval, 19
complete ordered field, 17
Complex Numbers, 15
composition of two continuous functions, 49
contiguous sets, 21
continuity at a point, 41
continuity on a set, 40, 41
cosine function, 160
Creeping Lemma, 23
critical point, 90
critical value, 90

D'Alembert's Ratio Test, 140
Definite Integral, 121
derivative, 70
differentiable, 70
differential coefficient, 70
domain of a function, 25

empty set, 8

Existence Theorem for Averages, 113
exponential function, 160

field, 15
First Fundamental Theorem of Calculus, 123
function, 24

graph of a function, 24

image of a set under a function, 25
infimum or greatest lower bound, 17
Integers, 13
integral, properties, 122
Integration by Parts, 126
Integration by Substitution, 125
Intermediate Value Theorem, 58
intersection of two sets, 8
interval, 19
Inverse Function rule, 86
inverse image of a set, 26
inverse of a function, 30
inverse to tangent function, 166

local maximum (minimum) point, 89
Lower Step Function, 115

maximum point, 87
maximum value, 88
Mean Value Theorem, 93
members of a set, 8
minimum point, 88
minimum value, 88
modulus or absolute value of a real number, 20

natural logarithm function, 169

INDEX

Natural Numbers, 13
neighbourhood of a point , 20

one to one function, 29
open interval, 19
ordered field, 16
ordered pairs of elements of sets, 10
ordered set, 16

polynomial function, 26, 145
power series, 146
product of two continuous functions, 47
product of two differentiable functions, 76
Product Rule, 86

quotient of two continuous functions, 51
quotient of two differentiable functions, 82
Quotient rule, 86

radius of convergence, 146
range of a function, 25
Ratio Test, 139
rational function, 29
Rational Numbers, 13
Real Numbers, 13
reciprocal of a differentiable function, 82
reciprocal of a continuous function, 49
Reciprocal rule, 86
relation on a set, 11
Rolle's Theorem, 92

second derivative, 72
Second Fundamental Theorem of Calculus, 123
sequence
    convergent, 133

divergent, 133
limit, 133
nth term, 132
series, 135
series sum, 135
set, 8
set difference of two sets, 8
sine function, 160
Step Function, 113
subset of a set, 8
sum of two continuous functions, 45
sum of two differentiable functions, 75
Sum rule, 86
supremum or least upper bound, 17

tangent function, 160, 165
The Contraction Property, 100
The Subdivision Property, 100
transitive relation, 12
triangle inequality, 20
trigonometric functions
    addition formulae, 162

unbounded interval, 19
union of two sets, 8

www.ingramcontent.com/pod-product-compliance
Lightning Source LLC
Chambersburg PA
CBHW030943180526
45163CB00002B/682